One Star by Day, Six Thousand by Night

Discovering the universe

Rod Somerville

First published by Rodney Somerville 2021

Typeset in Garamond by the publishers
Layout and design by Caroline Rich

Cover image courtesy of Geoffrey Wyatt

National Library of Australia
Cataloguing-in-Publication data

Creator: Somerville, Rodney, author.

Title: One Star by Day, Six Thousand by Night: Discovering the Universe / Rodney Somerville.

ISBN: 978-0-6450987-0-9 (paperback)
ISBN: 978-0-6480628-6-8 (e-book: epub)

Subjects: Astronomy, Stargazing, Telescopes, Cosmos, Stars, UFOs, Biography

To my family and friends,

without whom, none of this would have been possible.

Contents

PART THREE - TELESCOPES AND STUFF

Preface

A long time ago one of my brothers did something that changed my life. He gave me a book as a birthday present. He gave me *Second Foundation* by Isaac Asimov.

I was turning 11 at the time and was naturally curious about, well, everything, but until that moment I had no focus for my curiosity. Now, the third book of a trilogy was perhaps not the best one to begin with, but unknowingly my brother had picked one of the greatest science fiction stories ever written. And even though it was the last book of the series, it straight away ignited a lifelong passion for space and science.

Fifty years later, I still have the book and I cannot thank my brother enough for opening that doorway for me. Needless-to-say, after feverishly reading *Second Foundation* I raced out and bought the first two books of the series and devoured them just as quickly.

It was the right book given at the right time that sparked my imagination, but it doesn't have to be a book. Perhaps a trip to a museum, a particular TV series, or an enthusiastic teacher will make a subject come alive for you. But for me, reading *Second Foundation* was the moment when I knew what I wanted to do.

So, this book isn't about facts and figures or any particular topic.

My original copy of
Second Foundation by Isaac Asimov.

Instead, in the hope of inspiring you to start your own journey, I want to share with you some of the experiences I have had and things I have learnt as I travelled the fantastic path that took me to the stars.

Part One, 'A Personal Journey', describes the genesis of my love of the heavens and some of the adventures I had along the way.

In Part Two, 'Lessons Learnt, Knowledge Gained', I share some of the knowledge that has engaged and interested me as I explored the universe. It also contains a handful of straightforward observations and activities you can do for yourself to help get started in your own exploration of the universe.

As the name suggests, Part Three, 'Telescopes and Stuff', deals with telescopes, how to use one, what to look for when buying a telescope, and other useful tips.

I hope you enjoy this journey and along the way discover things that fire your imagination and get you to look and wonder at the universe found above and around us all.

Rod Somerville

Part One
A personal journey

CHAPTER 1

The journey begins

At the age of eight, I watched Apollo 11 land on the Moon. At the time I was certainly fascinated by it, but it wasn't the defining event for me that it was for a lot of people. Don't get me wrong, I was captivated, but it was just another exciting event in the world around me. But before we talk further about Apollo 11, let me digress a little and give you some background on how I ended up sitting on the floor watching perhaps the greatest achievement so far made by humankind.

In the 1960s, living in the suburb of Berowra Heights on the northern outskirts of Sydney was like being half in the city and half in the countryside. These days it is well within suburbia, but back then it was still lightly populated. So much so that just one primary school catered for everyone. But as the population grew, a second primary school opened and as luck would have it, right across the street from where I lived.

Complete with an abandoned house, the site was an old orchard and the location of endless exploration for myself and other children of the area. Of course, everything changed once the school opened. We lost a great place to play, but it did mean I no longer had to walk 1½ kilometres to get to school. I now only had to cross the road and jump the fence.

When the new school finally opened, for some reason, instead of simply having the relevant students start the day at the new school, we had to first go to the old school. Then, at lunchtime, we were marched as a group to our new place of learning. I can still remember feeling annoyed at having to walk 1½ kilometres to school, then walk 1½ kilometres to my new school, only to end up 30 metres from where I had started.

Only a few weeks after we had settled into our new school, complete with that new school smell, the entire student population was shuffled into the only classroom big enough to accommodate us all. There we sat, watching the school's one and only, state-of-the-art black and white television, waiting for one of the most significant events in human history, the landing of Apollo 11. There were no classes that day and we sat for hours watching the television, waiting for the moment. No one even wanted to go to the bathroom for fear of missing it. As it turned out, we sat there for a long time. We knew the astronauts had landed before school started at around 6 am, so understandably the teachers thought they would emerge sometime before lunch. However, unknown to anyone at the time, the astronauts were meant to rest for a few hours before getting out.

As you would expect though, when the Apollo 11 crew finally touched down they were far too excited to rest (who wouldn't be!), so Neil Armstrong decided they would exit straight away. But they had underestimated how long it would take to get into their spacesuits and getting ready ended up taking almost as long as their planned rest break. So when Neil finally stepped out it was now lunchtime and we were getting restless from sitting on the hard floor all morning watching nothing happen. Never-the-less, we were all still there when at 12:56 pm AEST on the 21st of July 1969 Neil Armstrong finally set foot on the Moon. Like everyone else, I will never forget that moment.

Even at the age of eight, I realised this was an enormous event, so I kept every newspaper article of the landing I could get my hands on. I filed them away, thinking that someday I would want them, and my young self was right. On the 50th anniversary of the Apollo 11 landing, I still had the articles and they had proved to be immensely useful as part of my professional life.

I should mention that as the years passed there were times when I thought about getting rid of them, especially each time I moved house. But every time I would say to myself, not yet. Now they have been laminated

LATE FINAL EXTRA

No 19,527 MONDAY, JULY 21, 1969 Five cents

THE SUN

ON THE MOON

• CITY: Showers. Colder tomorrow. • Lottery: New Jackpot No. 302, page 27 • Finance, page 28 • TV, page 37

MAN ON MOON

HISTORIC WORDS: "The Eagle has landed"

MAN'S centuries-old dream came true today at 6.18 a.m. (Sydney time) when the U.S. astronauts Armstrong and Aldrin dodged a huge crater and landed their lunar space vehicle gently on the rock-studded moon in a cloud of lunar dust. Full report. Pages 2, 3, 4 and 6.

FIRST STEP FIVE HRS. EARLY

THE incredible moment when man first sets foot on the moon will probably be brought forward by more than five hours today—to 11 a.m. Sydney time.

After an exhaustive systems check, the exhilarated astronauts Armstrong and Aldrin urged space mission control at Houston to abandon the scheduled program which would have put Armstrong on the moon at 4.21 this afternoon.

Full report, Page 3.

Front page of a newspaper kept
from the day Apollo 11 landed.

and should survive until long after I don't have any more use for them and it will be up to someone else to decide their fate.

For years after the first moon landing I read everything about the world around me that I could get my hands on. It was pre-internet days, so the only way to get information was through books, magazines and the occasional television show. Every so often, I was able to talk my father into taking me to the old Sydney Planetarium at the Museum of Applied Arts and Sciences in the centre of the city, complete with a life-sized mock-up of the Apollo Lunar Lander at the front entrance. I loved it, as I'm sure a lot of people did.

I also became an avid fan of science fiction stories and television shows. Over the years, I built up an impressive library of some of the greats in science fiction. Isaac Asimov, Arthur C Clarke, and Ray Bradbury being some of my favourites. Most of them are still sitting in the bookshelves behind me as I write this. There was no way I was ever going to part with them as I grew older. I was also a super fan of the television series, Lost in Space. I liked Star Trek, but for some reason Lost in Space was the one I couldn't miss each week. Looking at them today they are quite corny, but for a young boy so soon after the last Apollo mission, they were exciting.

Since this was well before the internet and the enormous amount of information it provides, this obsession with science fiction stories and television shows is what fuelled my hunger for all things science and pointed me towards my future. Looking back, I realise it would have been easy for my parents to squash this fascination as nothing useful, but instead, they supported and nurtured it, and for that, I am eternally grateful. I'm sure my father didn't like Star Trek and Lost in Space, but he still let me watch them whenever I could and sat there watching them with me.

As a related aside, 20 years later I had a colossal geek moment. While working in Central Australia conducting tours of the night sky, my evening constellation talk involved pointing out the Southern Cross and its two Pointer stars. I would mention that the Pointer furthest from the Cross is a star we should get to know a bit better as it is the closest star system to the Sun, Alpha Centauri. I would then say something along the lines of 'Dr Smith has a lot to answer for as they got lost trying to get to the closest star but somehow managed to visit dozens of others along the way'. Some people would get the reference. Most didn't. But one night, after making my Lost in Space comment, a gentleman came up, thanked me, and then told me he once worked as the set designer for the show. We spent the next half an hour being geeks and talking all things Lost in Space. I was in heaven.

But I digress.

As I said earlier, by the time I was 10 years old, I was naturally curious about everything around me. But at the point when my brother gave me Isaac Asimov's book, my attention focussed, and a lifelong passion for space was born.

A few days after getting the book, I read a passage near the end of *Second Foundation* and was so excited that I had to tell someone, so I told my soon to be sister-in-law, as she was the only other person in the room.

She was only 18 or 19 years old herself, but she indulged my enthusiasm and let me carry on as only an excited almost-teenager can. I have never forgotten that moment. It was a tiny thing, but it was crucial. Allowing me to carry on about something that must have been extremely uninteresting for her was a kindness I still appreciate, and it fed my curiosity and helped lead me to what I would become.

As it does for everyone, primary school eventually finished, and high school began. Going to high school in the 1970s was difficult, especially at an all boys school. When we started in Year 7 (or 1st Form as it was known back then), we would assemble in a circle every recess and lunch, all 200 of us, for protection. Being picked on and drawn into fights was just part of everyday school life back then. We quickly learnt that those on the outside were fair game to the older boys, but being in a circle meant those in the middle could enjoy a reasonably peaceful break. Each day the outside population would change, so everyone got a chance to enjoy lunchtime at some stage. These days, something like that would never be acceptable, but back then it was just the way it was. One positive outcome of this routine was that it did create a circle of close friends who remain so today.

I mention this because in 1973, our first year of high school, I convinced about eight of my new school friends to stay over one night to see a comet that was going to be visible, Comet Kohoutek. It would be my very first comet, and since it was advertised as the comet of the century, I was excited. The problem was that, according to the news, the comet was only going to be visible in the early hours of the morning. Being 12 at the time meant, of course, we just had to stay up all night.

The next problem was that the news, where we heard about the comet and got our information from, was very light on details about how to find it and what time to start looking. The time problem didn't faze us because we knew we were going to stay up all night, so whenever it was visible, we would be ready. But none of us knew the sky well enough to know where to look. I now know that the place where we searched for it was completely wrong. However, perhaps the most significant problem was that it didn't live up to expectations and was a lot fainter than predicted. Oh, for the internet back then!

In later life, this experience taught me that when giving details about comets and how to see them, I made sure to provide information that even the most astronomically untrained person could follow. Even today, when every detail about a comet is available on the internet, simple, straight -forward explanations are still needed to help people.

Having said that, one time I perhaps didn't explain myself as clearly as I could, or should, have. I was working in Sydney at a museum and we were experiencing an extended period of clear, stable weather and a comet was faintly visible to the west for a short period immediately after sunset. I can't remember the name of the comet as it wasn't particularly spectacular or well-known. One day, about lunchtime, long before it would be visible, a gentleman called to enquire about the comet. He thought he saw it the night before and wanted to check if that was what he had seen. He explained where and when he saw the object and I told him it could have been the comet but to have a look again that night. Comets tend to move little from night-to-night, especially the fainter, further from the Earth ones. If he saw it in the same position at the same time it would confirm it was the comet and not something else.

Anyone familiar with looking at the sky and comets in particular, will perhaps by now see my mistake. Because the initial phone call was in the middle of the day, I couldn't confirm his sighting by looking out the window. The next day he rang again, this time later in the day, saying he could see it right now and to thank me for the information. This time, however, I was able to look out the window. As soon as I did, I realised what he was looking at was not the comet but the condensation trail from a plane. I had not given it a thought that the weather at that time was conducive to planes creating trails and that at the same time every afternoon the same flight would take the same flight path, meaning each afternoon at the same time in the same location in the sky there would be a contrail. I immediately told him it wasn't the comet, but he would have none of that. He said I had told him if he saw it again the following night at the same time and the same location then it was the comet. No matter how much I apologised for my mistake, he refused to believe me. To this day he probably still thinks it was the comet he saw.

Over the next few years of high school, my interest in things astronomical ticked along. But by the time Year 10 (4th Form) arrived, and I had to select subjects for my final two years of school, I knew what I wanted to do but not what I needed to do to get me there. So, at the prompting of my parents, I rang the only place I could think of to ask for advice, Sydney Observatory. I was transferred to one of the astronomers and ended up discussing how to go about eventually getting a job in the field. By the time I hung up, I knew what subjects I needed to take, essentially as much Physics and Mathematics as I could. So that's what I did. Little did I know that six years later I would end up working there as a science educator with the astronomer I spoke to back in high school. I don't think he ever knew it was me.

In an interesting twist of coincidence, during Year 11 (5th Form) the school organised a trip out to Central Australia. This excursion began my enduring love for the Red Centre and I had no idea at the time just how important the Outback would turn out to be in my career as a science educator.

CHAPTER 2

Post high school

With high school over, it was time for university. As planned, I threw myself into as much mathematics and physics as I was able, intending to become an astrophysicist. However, while studying for my undergraduate degree at Macquarie University, I started to do some public education of science, in particular, astronomy and found I liked it. I wanted to do more. Before I could get too carried away, though, I had to finish my degrees in physics and maths, after which I started my Masters in Astrophysics.

As any university student knows, there is only so long your parents will support you financially. Consequently, as I entered my twenties, I was expected to start contributing to my own financial support. My parents were very good and never said it out loud, but I knew I had to start earning more than a few dollars to help support myself. On weekends I drove taxis all over Sydney, and let me tell you, that was an interesting job! I also worked casually in the evening as a science educator at Sydney Observatory. I did not know it at the time, but this was the start of my future career path.

The Observatory had only recently closed as a working facility, and even though it occupies one of the best locations in the city, rather than demolish such a historic and magnificent building, it was transferred to the Museum of Applied Arts and Sciences. That meant it was now a museum

Sydney Observatory.

and science centre. It was also why they advertised for casual science educators and the reason I now had another job.

I am fundamentally a shy person and find it challenging to engage with strangers, which may seem an odd thing to say as I now had a job talking to continually changing groups of people. But astronomy, and science in general, was my passion and talking about it came easily, so my inherent shyness was not a problem.

The first couple of years working there, I think, finally determined my life. The people I worked with were great, as was the general atmosphere of the place, and converting a working museum full of beautiful old astronomical equipment into a functioning science museum was exciting. I realised that being a science communicator was what I wanted to be, so

Halley's Comet taken from Maree, South Australia.

The southern end of the Birdsville Track at Maree, South Australia.

I left university and went full time explaining science to the public as a science educator.

Since it was still a transitional phase for the observatory, in the beginning it took a while to get used to showing visitors around what was primarily a working observatory, minus the working part. The building had the feeling that everyone just stood up and walked out one day, leaving everything behind and giving the building an eerie feel to it, especially at night. When the museum installed some yellow spotlights to illuminate the outside of the building, it didn't help to alleviate this impression. Instead, it just gave the whole site a sickly, slightly horror movie-ish appearance. Arriving at night, I would half expect to see a ghostly figure looking out of the top floor window.

I wasn't the only one who had this impression of the building. As part of the evening tour, each guide would take their group of visitors down to the basement, where we had some instruments and atomic clocks stored. One of the other guides was so spooked by the ambience within the building that whenever it came time for her to go down to the basement, she wouldn't do it. We had to swap groups at that point, and I would end up going downstairs twice each night we worked together. That such a rational, intelligent person would have such a fear always amused us, but we still adored her. To us, it was just one of her quirks.

In 1986 I was involved with my first major astronomical event. Halley's Comet had returned, and the world was abuzz with excitement. Remembering Comet Kohoutek a decade earlier, I had learnt my lesson and was determined to make sure everyone knew where and when and what to look for. We all knew Halley's Comet was never going to be as spectacular as it was the last time it had appeared in 1910, but that didn't stop the media from going berserk. It did, however, teach me my first couple of working life media lessons. Firstly, it doesn't matter what you say, the media will edit whatever it is to the point where it is totally out of context. Secondly, they will only use seconds of the interview. To avoid these problems, you need to develop the art of getting your point across in five second soundbites.

Even though I knew what to expect from the comet, Halley was my first major one since Kohoutek and I was excited. But at the time a lot of people were disappointed. They were expecting the apparition hyped-up by the media, and it was never going to be that. The thing to keep in mind, and I said it many times while it was visible, is how many bright comets have you seen? Since Halley, there has been only a handful bright enough to see using only your eyes. Halley may not have been as big as some people expected, but it was easily visible from my backyard in a light-polluted suburb of Sydney.

I was also fortunate to have the opportunity to see Halley from an extremely dark site. Because of the Halley hype, a tour company decided to market one of their camping tours to Central Australia explicitly to view Halley's Comet. They needed a guide with specialist knowledge to give talks on the comet and the night sky as we travelled around the country. Since the tour company did not have their own experts, they rang work and asked if anyone would be interested. I was the first to put my hand up, so I got to go. It turns out they had such a good response that they ran two tours side by side. A friend became the expert on that bus, while I was the expert on mine. We both got a free, two week trip to Central Australia showing people the night sky, which we both loved doing. Talk about a great deal!

As a bonus, while we were at the remote South Australian town of Maree (at the southern end of the Birdsville Track), we were treated to a partial solar eclipse. My friend and I had fun showing everyone how to look at the eclipse safely in one of the most remote places in Australia.

I had enjoyed the trip to The Centre so much that I wondered if I wouldn't mind doing this as a job. But as much as I enjoyed the touring, I think it was the opportunity to stargaze from a truly dark place that attracted me. So, as soon as we returned, I applied and got a casual job as a guide travelling all over Australia, which I ended up doing multiple times. Somehow over the next two years, I manage to fit in these trips with my job at the observatory. It wasn't easy, and two years of squeezing both jobs in were enough, but I wouldn't change those years one bit as I thoroughly enjoyed every minute. Having seen most of the country, I now appreciate just how blessed we are with natural beauty and how all Australians should see more of their own country before venturing overseas.

Once back in Sydney I continued working at the observatory (while somehow fitting in the touring job) but also started to conduct adult education courses in astronomy for several different colleges. These taught me more presentation skills and gave me the confidence to prepare

Myself as a tour guide somewhere on the Tanami Track, Western Australia.

and conduct talks for quite large groups of people. I was still shy about speaking to groups, but after spending the half-hour before the talk pacing nervously, as soon as I started, I relaxed and enjoyed myself. Hopefully, the people attending did as well.

From presenting the courses and running the evening telescope tours, I learned that people were interested in everything to do with space. Most had never looked through a telescope before, so anything they did see was exciting. Most didn't know much about astronomy and how to get started in the field, and they were hungry to hear someone tell them about it. I found I was best able to get across my sense of excitement and impart information if I treated it as a conversation rather than a lecture. People weren't looking for endless facts and figures. They wanted a relaxed, enjoyable time. Some facts here and there were a bonus and I tried to sneak in the learning without them noticing.

With each adult education course I presented, I would organise a trip out of the city to use a telescope in a dark location. In case the weather didn't co-operate, I tried to hold the viewing near a major attraction, so if it wasn't possible to see the stars, we still had something to do. At the time a friend was working at the Parkes Radiotelescope in the Central West of New South Wales, so for a number of the courses we would go out to the town of Parkes, have a tour of the telescope, do some stargazing that night then come home the following day. It made for a pleasant weekend away.

The weather was usually kind to us, but on one memorable trip, we were stargazing under crystal clear skies when, after looking at about four objects, someone pointed out that the clouds were moving in quickly and they had seen lightning. I thought we had just enough time to see a fifth object before the storm arrived when suddenly, it started to rain. Fortunately, I had developed a habit of always having large plastic bags in my kit. Don't ask me why I started this habit, but I'm glad I did. I quickly threw one over the telescope and went to get my van. By the time I backed the van, with its tailgate open, up to the telescope to provide some protection and started to pull the telescope apart, it was bucketing down. Since that night, whenever I go stargazing, I always make sure to have the plastic bags with me just in case. Luckily, they have only rarely been necessary.

These occasional trips to darker skies in rural New South Wales always reminded me of the wide-open spaces and magnificent view of the stars visible from the centre of the country. Thankfully, I would soon be heading into the Outback once again.

CHAPTER 3

A Central Australian adventure

In 1990 the Ayers Rock Resort, located beside the world's largest monolith, Uluru in Central Australia, advertised for someone to set up an observatory and conduct evening tours of the night sky for their patrons. With the help of a few others, a close friend and I decided to apply. I had been to the resort a dozen times before with my guiding job, so I knew what to expect when I was successful. Or at least I thought I did.

Having a captive audience and crystal clear dark skies seemed the perfect opportunity to expand my love of the universe. It turns out I was right about the audience but unfortunately not so much about the clear sky. In 1990 the internet didn't exist, so we had to rely on information about yearly weather patterns provided by the resort and the Bureau of Meteorology. Unfortunately, once out there, we discovered that their definition and our definition of clear skies weren't quite the same, but we will come back to that shortly.

For our contract submission, we photographed our equipment one sunny weekend from the top of Observatory Hill in Sydney as we figured it would look better to have the city and harbour in the background rather than a blank wall. The photos turned out wonderfully, but shortly after we took the last picture, we noticed there were thunderstorms in the distance surrounding the city. So, before packing up the equipment, we indulged ourselves in a little bit of storm watching. This turned out to be a significant learning moment for all of us.

There were three storms: one to the north, one to the west and one to the south. As an understatement, the thunder and lightning from each were spectacular, and we lost track of time admiring the show. However, the next thing we knew, all three storms were rapidly converging onto the city. Broken from the spell of the lightning display we madly started to pack up the equipment but had left it a fraction too late. We managed to get the smaller pieces inside in time, but at the very last moment, before armageddon struck in the form of a perfect storm, all we could do was, once again, desperately throw large plastic bags over the telescopes. The heavens then opened up in one of the most violent storms I've ever seen. We couldn't leave the gear out in the rain so, while they were still under the plastic bags, we pulled the telescopes apart and carefully transported them bit by bit to safety undercover, where we could then pack them away properly. Fortunately, none of the equipment was damaged, but I can't say the same about us. I will never forget standing in the wild storm, completely soaked, pulling the telescopes apart. The Parkes incident had not taught me to pack things away at the first hint of rain, but this day I did learn my lesson.

Anyway, we eventually submitted our proposal, and it was good enough to get us an interview and trial presentation. The resort would pay our travel and equipment transportation expenses, and we would get to show them what we could do. Initially, three of us went out: me, my business partner and our technical expert. I call him our technical expert, but he was, in fact, a friend who was just more technically inclined than we were. Once out there, we had a day to set up and conduct an evening for about a dozen of the resort's management.

The following day we were to fly home, but in the morning we did something that probably swayed their decision in our favour. The resort had paid for our equipment to be shipped out and, of course, was going to pay for it to be shipped back to Sydney. From our perspective though, we thought if they chose us we could avoid having our delicate gear making a 5,000 kilometre round trip by road if we left the equipment at the resort until they made their decision. If they chose us, we already had our equipment on-site, and if we weren't successful, all we had done was delay slightly what the resort was going to do anyway. We were thinking purely about moving around a lot of equipment, but in hindsight, I think it may have been the thing that influenced their decision towards us.

As you may have guessed, we did get the job, and that meant finishing up in Sydney and getting ready to move 2,000 kilometres away. I say 2,000 kilometres, but that's in a straight line. By road, it was more like 3,500 kilometres and a long way to drive with a van full of possessions. It took us three days to get there but we made it with two days to spare before we were contracted to start operating.

Since this was a brand new venture for the resort, when we arrived, there was no specific custom-made site for an observatory. That would come about six months later. In the interim, the plan was to use the existing amphitheatre as our 'observatory'. This was not ideal, as a large shade sail and trees blocked a lot of our view, but until something more suitable could be constructed, it served our needs well enough. But unfortunately, there was nowhere in the amphitheatre to store our gear. So, every night we had to carry the equipment from a motel room, across the road, down the stairs of the amphitheatre, onto the stage and set it up. At the end of

Our first observatory in the resort amphitheatre.
The closed-in space storing our equipment is at the back, underneath the screen.

the night, around midnight, we repeated the process in reverse. We were more than a little tired, but at least we were becoming fitter.

After a couple of months of this arduous regime, the resort built a temporary, closed-in area at the back of the amphitheatre stage so we could leave the equipment there. This saved a lot of effort but also brought with it a nasty surprise. Anyone who has been to the Australian Outback knows it is a truly spectacular place that is also home to some troublesome creatures. Within days of storing our equipment at the back of the stage, we found each evening that our gear would be covered in Redback spiders. They are not the most dangerous spider in the world, but they are still nasty.

Growing up in Sydney with the deadliest spider, the Funnel Web, meant a few dozen Redbacks each night didn't bother me. I remember my grandmother's back yard riddled with Funnel Web holes, and whenever we visited, we knew there were safe areas and paths to follow. Venture outside these areas, and you had to be wary. Very wary. Years after my grandmother died, I drove past her house and noticed it had been demolished and apartments built on the site. I bet the workmen that built the apartments got a huge surprise when they dug up all those angry Funnel Web spiders! Living near the edge of the bush on the outskirts of Sydney, we also occasionally found aggressive Funnel Webs inside the house. But being aware of spiders was just part of growing up, so when it came to Redbacks all over our equipment at the resort, the solution was not to panic or use copious amounts of spray but to leave a rock near the door. Each night we would pick up the rock, open the door and before we touched any piece of equipment squash any spiders that happened to be on it. The method worked well.

Things would occasionally take an exciting turn when we found a snake amongst our equipment, but we survived those moments as well. These weren't harmless either. They were Western Brown snakes, among the deadliest in the world, so some caution had to be encouraged.

After arriving at the resort and before our first official session, we figured we had better do a run through and familiarise ourselves with the Central Australian sky. Back in Sydney, it was easy to locate the main constellations as you could only see the brighter stars. That meant only a limited number were visible and identifying the constellation shape was relatively easy. But when we looked up that first night in the Centre, there were so many stars visible, from horizon to horizon, that it was difficult even to locate the Southern Cross, one of the brightest and most recognisable constellations of all. This caused us a bit of concern. If we struggled with the Southern Cross, how could we conduct an hour and a half tour only one day from then?

We spent the next few hours working out how to locate and identify just four constellations, which we figured was sufficient to tell stories around and point out objects within their boundaries. The plan was to start simple and expand as time progressed. As these four slowly disappeared, we would add newfound constellations and objects to the repertoire. By the time a year had gone by we were able to point out maybe 70 of the 88 constellations up there. The only ones missing were those too close to the North Celestial Pole for us to see.

Eventually, the time came when we moved out of the amphitheatre and onto a purpose-built decking and storage area. The resort is laid out in a rough ring shape, with only dunes and walking tracks in the middle, and it was within the central dunes we built our observatory deck. We put it there because the dunes blocked some of the light from the resort, plus the site offered easy access from any location. It also gave the ambience of being all alone in the desert, even though the resort buildings were only 100 metres away, behind the dunes.

The Ayers Rock Resort and town of Yulara from the air.

Our new observatory nestled in among the dunes... and the snakes, scorpions and spiders.

View of the observatory with Uluru in the distance.

The main observation deck with the 20 cm SCT, 28 cm SCT, and 25 cm astrograph.

A close-up of the 25 cm astrograph on the main deck of the observatory.

Something we didn't anticipate when designing the observatory was the number of amateur astronomers who said we should have built domes for the telescopes, rather than open decks. We couldn't understand this as we had specifically opted for an open deck plan in order to be under one of the great natural wonders, a truly dark night sky. The last thing we wanted was to be inside a dome. It may have been better for the telescopes, although that was debatable, but from an experiential perspective, being under the open sky was the only way to go.

Another thing that took me by surprise occurred when construction of our new platform was nearing completion. Talks turned to the signage to be placed at the start of each track leading across the dunes to our new home. The resort wanted to know what we wanted on the wooden signs. We said our name was Southern Skies Observatory so, of course, that's what should be on the sign. The resort, however, had the idea that the word 'observatory' meant a bird or animal observation deck and therefore our name would confuse people and no one would know we were looking at the stars. I can understand how the word observatory can be used to reference a bird viewing platform, but I thought everyone knew that an observatory was what you called a place with telescopes. But apparently not. It took a while, but we eventually got the signs we wanted. I still have those signs, proudly hanging in my back yard as it was part of our deal when we left that we got to keep them.

With a proper storeroom for our equipment, spiders and snakes were now less of a concern and setting up the equipment was a lot quicker and easier. This was a good thing, because in the desert during winter the temperature could drop from 20°C half an hour before sunset to -5°C ten minutes after sunset. If you were outside at this time, you could feel your lips and face dry out from the cold minute by minute as the temperature dropped. On these nights, people would come along wrapped in blankets, and I didn't blame them. I wished I could have had a blanket as well.

With a permanent home, we could now also do some solar viewing during the afternoon, which was an excellent way to advertise and upsell visitors to the evening tours. It was enjoyable for everyone, especially myself, as looking safely at the Sun is something everyone should do at some time in their life. It is always fascinating.

One of the signs at the start of a path leading to our observatory at the Ayers Rock Resort.

The same sign is now hanging in my backyard.

Solar viewing on the main deck with our Japanese guide.

Of course, solar viewing is all well and good during the winter months. But summer in the desert is brutal. The first summer we were there the temperature in the shade on my back porch in the early afternoon was over 50°C for 10 days in a row. Thankfully, at night it would ease back to a balmy 38°C. Essentially, I turned the air-conditioning on in October and didn't turn it off until April.

Now, with these sort of temperatures during summer, you have to take precautions, especially if you want to stand out in the sun for a couple of hours each afternoon looking through a telescope. Lots of sunscreen, long sleeve shirts, hats and shade from an umbrella were mandatory. Even so, there were days where we had to stop early because the heat was just too much. Australians know that to be out in this sort of heat unprotected for even just an hour can cause death, but many tourists didn't seem to understand. There were always some found sunbaking beside the pool mid-afternoon in 50 degree temperatures.

Looking at the Sun during the day is great, but looking during the evening was why we were there. If you have never been to a really dark place in the country, do yourself a favour and go. Faint stars are visible right down to the horizon. The Milky Way leaps out at you. There are so many stars it seems you couldn't count them all (although, theoretically the figure is only about 3,000 individual stars visible to the eye at any one time, but I think that bit of information spoils the spectacle). And faint objects viewed through the telescopes stand there in all their glory.

During my time in Central Australia, I managed to see so many objects that I had only ever read about before. In particular, those visible with just binoculars and my eyes. While my partner was conducting his part of the evening tour and I was momentarily free, my favourite pastime was to lie on my back, off to the side, and stare at the incredible night sky. I know it isn't possible, but I swear I could see the Milky Way in three dimensions during these times.

I also came to appreciate just how much the Moon affects our view. From a town or city with light pollution, the effect the Moon's light has on your view isn't noticeable, as artificial lighting interferes as much, if not more. But with no artificial lighting, the Moon's light is dramatic.

Another thing I came to appreciate was the weather. What I consider to be cloudy weather, and what the majority of people call cloudy weather is not necessarily the same. Most people thought it clear as long as it wasn't raining. This is not a great definition when trying to use a telescope.

But cloudy weather in Central Australia does have an upside, even for an astronomer. I have always loved thunderstorms, and the storms in the desert are spectacularly amazing with most consisting of a single cloud delivering a fantastic lightning display. If the storm passed over you, the rain was torrential but would stop quickly once the cloud moved on. Since the desert is mostly flat, this torrential downpour of rain has nowhere to go, so the resort would temporarily flood until the water soaked into the ground. These were great opportunities to race outside with a cloth and wash my car. Water is precious out there, and you weren't allowed to wash your car any other time so I couldn't let the opportunity pass.

The other thing with storms was you could always tell who was a local when it rained. They were the ones heading toward Uluru while it was raining to see water cascading off it. An incredible sight to behold when it

Lightning over a distant Uluru.

Water cascading off Uluru during a storm.

happens. Tourists were the ones heading from Uluru to the resort because they didn't want to get wet.

Often storms would pass by off to the side. During these times, we had crystal clear skies overhead and the storm's fantastic display in the distance. Occasionally we would just have to stop and watch the storm for a few minutes before carrying on with our stargazing.

Of course, there were also times when it was completely cloudy, and desert weather is notorious for changing within minutes. On one night the weather was clear, no breeze, and balmy. People arrived for the evening tour and, because of the weather, we had a big crowd, around 50 people. I had finished taking the admission fees and mustered everyone together to introduce myself when I noticed a cloud front moving in quickly. Within two minutes, it had completely covered the entire sky. I hadn't even finished welcoming everyone before I had to call the night off and hand the money back. Thankfully it was dark so people couldn't see me crying as I did so!

Looking for a positive side to these cloudy nights, I did come across one intriguing thing. Another regular tour we provided was to give a star talk for a tour company at the end of their walking tour around Kata Tjuta, an impressive natural formation about 30 kilometres from Uluru. That meant each night one of us would drive out to meet the tour for dinner and conduct the star talk, while the other person started the main tour at the observatory back at the resort. When we finished at Kata Tjuta, we would drive back and hopefully arrive in time to help with the telescope part of the main tour. It worked well enough, but you didn't want to be in a hurry.

A few times, when I was running late and perhaps driving a little too fast, I would come across kangaroos or camels in the middle of the road. Having to pull up from 140 km/hr in the space of a few metres is more than enough to get your heart racing. Kangaroos are impressive enough, but a camel whose stomach is the same height as your windshield is another level of awesome. Apart from the wildlife, one of the best things about the night time drive was that you were kilometres from any source of light. If there was complete cloud cover, so not even stars visible, the night would be pitch black. A few times as I was driving, I would turn off my car headlights to see what it was like and found I could not see a metre in front of the car. It was total blackness. Animals were not the only way to get your heart rate up on that drive.

Before I leave the weather, I have to mention sand storms. Frequent and massive in extent, if I saw one coming I had to rush around putting towels against the gaps under doors, around windows and make sure all telescope equipment was tightly packed up. Even so, the sand and dust got everywhere. If your clothes and shoes weren't red to begin with, they certainly were after a few sand storms.

If you are a space buff, red sand and rocks and dust storms may sound familiar. Take away the spinifex and mulga bushes (plus a few other bits of flora and fauna), and you would find Central Australia and Mars look very similar, and for the same reason. The sand and rocks in both places have a small amount of iron oxide (also known as rust) in them, and that gives them their reddish appearance.

Incidentally, I love the desert, but here's a tip. If you ever visit spinifex country DO NOT touch the spiky grass, it is nasty. The spikes have needle-like tips that break off in the skin and cause their victim no end of grief for days.

Contending with the natural world is one thing, but running your own business is tough. In the beginning, you don't have resources or money, and your time is 100 per cent devoted to establishing yourself. If it is

Moonrise over Kata Tjuta.

Dune with red sand and spinifex on the western edge of the Simpson Desert at Old Andado Station, east of Alice Springs.

Remains of the old Ghan Railway track in South Australia.

Moonrise over a glowing, red Uluru.

your first business, it is doubly hard. Moving out to Central Australia and establishing a profitable public observatory in an already thriving resort was both a blessing and a challenge. It was a blessing because they already had marketing systems in place, plus an established visitor base. It was a challenge for the same reasons. It meant we had to slot into this existing system immediately while establishing our business.

At times it was difficult. People on the east coast would call us at 7 am (our time) asking for information and decisions while at the end of the day we wouldn't finish work until around midnight. Working every day and every night like this would eventually take its toll. After about three months one of us would crack and either have to go for a few days break to the town of Alice Springs (about 450 kilometres away) or fly back to Sydney for a week. That meant the other person had to do double time while the other was away. Because of this, once one of us returned, the other then usually needed a few days off. After another three months, the cycle would repeat itself. During our time in the Centre, my father unfortunately passed away unexpectedly, and I had to fly back to the east coast for about two weeks. How my business partner survived doing everything for those two weeks I will never know, but I am eternally grateful to him.

Aside from the stress of continuously working, living in the Centre was great. The lifestyle and people were relaxed and friendly, the stargazing was amazing, and the experiences were incredible. We occasionally managed to sneak in some sightseeing, going to places only the locals knew, plus we got to meet some wonderful visitors.

During the time we were out there, the resort had a turnover of 7,000 visitors per night during peak times and around 5,000 during off-peak times. For us, that meant a never-ending stream of new people. It blew our minds the first time a visitor from Europe said that someone they met while backpacking in Germany had told them they must come to see us. We were internationally famous! At least that's how we ran it in our heads. Truth be told, it may have only happened a few times, but as far as we were concerned our fame had spread far and wide.

Our fame wasn't restricted to just word of mouth. While we were there, it was a never-ending stream of radio, newspaper and magazine interviews, as well as appearing in a diverse range of publications across the country. Whenever a TV crew visited the resort, we were on the interview list. We did regular radio spots on local radio stations and frequently conducted special tours to sell our product to visiting VIPs.

With such a high turnover of people at the resort, and being the experts in all things to do with space, eventually we had people with 'unique' views seek us out to discuss their thoughts.

Some people claimed to have aliens living under their beds or been abducted by UFOs. It seems they wanted to know if we had seen anything. I guess they just wanted us to validate their ideas, but as we weren't friends with any aliens or had seen any spaceships, we couldn't do that. There were also groups of people that, and I hesitate to use the word 'cult', held particular belief systems that I could never understand. I remember one in particular quite well.

Uluru is understandably a place of immense cultural history. In a flat landscape, the monolith stands out and immediately draws your attention, so it is no wonder it holds a special interest to anyone who sees it. Now, you may have heard of something called 'ley lines' that supposedly crisscross the globe. These are meant to be lines of energy, and where they cross is particularly important and influential. Apparently, these crossing points are where we find cultural artefacts, such as Stonehenge and the pyramids

According to this edition of *Cleo* magazine, the observatory was Number 1 in the Romantic Getaway list.

Country Style magazine article about the observatory.

Article about the observatory in the Japanese astronomy magazine *Tenmon Guide*.

Me being interviewed by a TV crew not long after we started.

Waiting to be interviewed in the early morning sub-zero temperature on top of a dune.

of Egypt. There is also one of these crossing points under Uluru, or so I am told. I have even heard that Uluru is hollow with a spaceship hiding inside, but I don't think I should believe everything I hear.

Anyway, early in November of 1991, a group who followed a mystic called Solaris decided to meet at Uluru and channel its energy. Their premise was that the numbers 11:11 held special significance (the reason given to us was 'How many times have you looked at your clock, and it read 11:11?') and Solaris had convinced them that the time had come to use this. What they planned to do was get 144,000 people (again, apparently a significant number) to surround Uluru while holding hands on the first day of the eleventh month of 1991 (i.e. 1/11/1991 or 1111991, don't ask me what happens with the 991 at the end) and meditate for 24 hours. They had calculated that they needed 144,000 people, although walking around the base of Uluru is just under 10 kilometres so that would mean each person had only about seven centimetres in which to stand. Assuming a more practical spacing, if we allow half a metre for each person that means their circle of believers would be 72 kilometres around. That would place each person about 11½ kilometres from the rock, which seems to me like it would be too far away to suck up any energy from the ley line. The Solaris believers apparently also did not know that the national park closed at sunset and no one was allowed to be in there overnight. As it turns out, only a few dozen people showed up, so it wasn't an issue. Their plan may not have been successful, but they did provide us with a couple of days of interesting conversation as they tried to convert us.

Working at the Ayers Rock Resort taught me some valuable lessons. I learnt how to manage and entertain people, what it takes to run a business and the time commitment required, and how to make decisions. For just those lessons I would say my time in the Centre was invaluable, with the astronomy and people wonderful bonuses. But then our contract ran out, and it was time to move back to Sydney. The trip was almost uneventful, except for a premonition.

A couple of nights before I had to drive my van back, my business partner had a dream. He dreamt that when I got to the Erldunda Roadhouse at the intersection of the Lassiter and Stuart Highways (drive 200 kilometres due east from Uluru and at the end of the road turn right), I would have an accident. Consequently, he decided to fly back and leave me to drive the 3,500 kilometres in three days by myself. As expected, I didn't have an accident at the intersection, but my friend still maintains that's why he didn't want to do the drive. Thanks mate.

The drive itself was uneventful except that it was long. The first night I stopped at a roadhouse, the only one for about 300 kilometres, but they were full. Since I was too tired to continue, I decided to stay in the roadhouse carpark and sleep in the front seat of the van, the back being full of gear. Unfortunately, it was uncomfortable and I didn't sleep well. That meant I was awake early, and with nothing else to do at the roadhouse, I left. Some advice about driving around the outback, put a little bit more thought into it than I did at that moment. I left before the petrol station opened, but I figured I had enough fuel to get to the next town, Woomera, the old Australian rocket facility. I arrived at Woomera at 8 am, but its petrol station didn't open till about 10 am. Keen to keep going, for some unknown reason, I decided to push on towards Adelaide rather than wait the two hours. For the last 100 kilometres, the gauge on my petrol tank was below the empty line, and all I could think about was what would I do if I ran out of fuel. Would I leave the valuable telescopes and gear alone for hours while I hitched a lift to get some petrol? Would I stay with the car and hope someone would stop who had some to spare? I made it that time, but I have never let my tank run that low again.

CHAPTER 4

Return to Sydney

Back in Sydney, the first thing I noticed was how light-polluted the sky was. Growing up I never thought about it, but now that I had spent years under a gloriously dark night sky, I realised all I had was just a handful of stars visible.

Even though I was sharing a house with friends, my first priority was to get a job, as my savings would not support me for long. So, a couple of days after I returned, I went into the city to Sydney Observatory and, given my experience and that I still knew everyone, they gave me a casual position straight away. In what turned out to be a bit of great timing, it was only a couple of weeks before my old job became available again. I applied and once again found myself working full time.

In the following years at Sydney Observatory, there were the usual school tours, evening tours and adult education courses, but the observatory had now been a science centre for several years and things were expanding. As someone who had been there from the start and possessed a wealth of experience and knowledge, I was closely involved in developing programs and events. It was also one of my tasks to deal with the media.

Over the years, I have been interviewed by TV, radio and print media about all manner of things astronomical. Anything from my thoughts on UFO sightings, to upcoming events held at the Observatory, to explanations of astronomical phenomenon. I have to admit I didn't enjoy the TV interviews, as they tend to edit more severely than other media. Quite often, I came across as not making a lot of sense. Of course, it wasn't just me. They did it to everyone. I did, however, enjoy the radio and print interviews, as I could have a conversation with the reporter and lead the discussion in the direction I wanted it to go. Occasionally there were more extended, fun interviews with the media. I have appeared in print in an Australian Geographic article, been interviewed for a domestic in-flight magazine, conducted a recorded conversation for in-flight QANTAS radio and, perhaps my favourite, did a weekly series about the planets on ABC Radio. Each week I would talk for five minutes on the mythology, facts and figures for a different planet. It was a pity there were only the Sun, Moon and nine planets (Pluto was still a planet back then) to talk about.

In the beginning, the observatory had just one evening tour, held on a Wednesday, but by the mid-nineties, we had built the observatory up to the point where we were running two tours every night of the week, all year round. We had also started school holiday programs, which were specifically for children during the day, but inclusive of adults in the evening. We also started holding special viewing events of an evening. It would be an understatement to say we became creative with our marketing since we were fundamentally doing the same thing each time. However, the programs did manage to create interest and entice people to come and visit the observatory.

Our daytime school programs started at just a couple per week but were eventually built up to two every weekday morning during school terms. Public opening hours were initially 2 pm to 5 pm during the week and 10 am to 5 pm on weekends. However, before too long, we had extended them to 10 am to 5 pm every day of the week. We ran regular tours of the telescopes, talking about the instruments themselves and then using them to look at planets and stars during the daytime. If it happened to be cloudy, our go-to option was an object not quite as celestial, the city. The Sydney Harbour Bridge to the north, traffic on a bridge to the west, a lighthouse to the east, or a city building to the south were favourites. They were far enough away to demonstrate what the telescopes were capable of, but close enough to compare with the view we could see with our eyes.

View of the Sydney Harbour Bridge from the Southern Dome of Sydney Observatory.

View of city buildings from the Northern Dome of Sydney Observatory.

Like any major city, the traffic in Sydney is appalling and evening peak hour lasts until well after sunset. I lived out west, near Parramatta, and the last thing I wanted to do was sit in traffic trying to get home. So, before leaving, I would use the telescope to check the traffic on the bridge heading west. If it were light, I would go. If it was heavy, I would wait an hour. Either way, I got home at about the same time, so I figured it was better to sit at work than in traffic. It may not have been what the telescope was designed for, but I look at it as an inventive use of a handy tool.

One of the most satisfying aspects of working in a science centre is getting to meet and interact with people of all ages. Most were friendly and keen to listen and learn. But for me, the children who dragged their parents along because they were super keen and wanted to know everything they could about space are what made it all worthwhile. In the age before the internet, visiting museums and science centres was often the only way kids could discover things. I have many great memories of talking with children and parents about space, trying to answer their questions.

A child's enthusiasm and appreciation were more than enough to make my day worthwhile, but it was when they sent a letter expressing their gratitude that I would realise how much of a positive impact I could have. Encouraging their eagerness for an hour could make an impression that lasted a lifetime, and all it cost me was a bit of time. Over the years, I have thankfully received a number of these letters, and I still have them to this day.

It was a busy and productive time with special events sprinkled in amongst the daily routine. The observatory had made a conscious decision not to do a lot of outreach programs, despite constant requests, as we wanted people to visit the museum instead. We had a few portable telescopes, so it could have been possible, but we had decided it was better to have people come to us and use the larger ones in the domes or the smaller ones in the grounds. We did, however, do the occasional outreach for exceptional events and publicity purposes. Two exceptional events stand out in my memory.

The first was held at the amusement park on Sydney Harbour's northern foreshore, just under the Harbour Bridge. Luna Park had been there since long before I was born, and I had spent many days there in my youth. It was a fantastic place in a fantastic location. Before upgrades were made (I think wrecking the experience) the rollercoaster used to go out over the water of Sydney Harbour. Talk about a great ride. Anyway, one particular night in August 1995, Luna Park organised a special event and called it 'Festival of the Luna'. The observatory provided a telescope and some people, including myself, to operate it with the instruction of placing a particular emphasis on viewing the Moon. The festival only went for one night, but the weather was perfect, and up and until then, I don't think I had ever spoken to so many people in such a short period. We had a great time, made even better by receiving free passes to use at the park as thanks for being there.

The second event was at the Sydney Opera House in June 1996, where they were hosting a play about the life of Galileo. I guess as a gimmick they wanted telescopes on the Opera House forecourt for patrons to look through before, after, and at intermission in the play. I suspect it was more good luck than good judgement, but Jupiter was high in the sky at the time, so it was a perfect opportunity to view the planet and talk about how Galileo was the first to see Jupiter's largest moons. We were there over five nights (thankfully with clear weather the whole time) with two telescopes each night, and as thanks, everyone who worked was given tickets to see the play on the final night. It was a great play in a great location and we all thoroughly enjoyed the week long experience. After all, who wouldn't enjoy showing people through a telescope on the Sydney Opera House forecourt jutting into Sydney Harbour.

On the first night, there was a moment that still makes me chuckle when I think about it. The Opera House had invited several VIPs, including a world-famous astronomer. We had already shown a few people through the telescope before the play started, so we knew precisely where Jupiter was. During intermission, the astronomer and his friends came out to have a look and as they were looking through the telescope and listening to our explanation, one member of the group asked where Jupiter was in the sky. I pointed at the brightest object visible and said there it is. Even though I had been pointing the telescope at it for a few hours, the astronomer contradicted me! He told his friend I was wrong and that it wasn't Jupiter at all. I was sure which point of light was Jupiter, so feeling baffled and flustered, I started to say please have a look at where the telescope was pointing, but he stuck to his story, insisting I didn't know what I was talking about. When intermission was over, they went back inside, leaving us momentarily alone, and we had a good laugh about it.

This wasn't the only time I have met professional astronomers that don't know their way around the night sky. It seems if you want to know details about an object, ask a professional astronomer, but if you want to locate something, ask an amateur astronomer. Of course, like all rules, there are exceptions.

Without a doubt, however, the best publicity outreaches we did were at Circular Quay on the edge of Sydney Harbour. For four hours each sunny Saturday and Sunday over summer, we set up a telescope on the grassy area on the western side of the Quay and used a solar filter to look at the Sun safely. The observatory is at the top of the hill above The Rocks area, while Circular Quay is on the harbour's edge beside The Rocks. By setting up at the Quay, we could do our PR exercise and suggest people walk up the hill and visit the Observatory. I can think of very few better ways to spend a workday than being in the sunshine beside Sydney Harbour watching ferries come and go while talking to people as they peered through a telescope at the Sun.

Getting people to look through a telescope directly at the Sun requires a certain amount of trust. A lot of people weren't merely going to take our word that it was safe, so we soon learnt we had to show it was safe by looking through it ourselves. It was during one of these demonstration times that I had perhaps my best experience using a telescope. At that time the Sun was near the peak of its activity, so there were lots of prominences and flares visible on its surface. As I was looking through the telescope, I saw a truly spectacular flare occur on the edge of the solar disc. It grew to be about half the diameter of the Sun in size, making it around 500,000 kilometres high. That is impressive, but even more so is that it went from start to finish in the space of just 10 minutes, visibly changing shape and size as I watched in wonder. This was the first, and so far only, prominence I have seen change shape so dramatically in front of my eyes. I was awestruck. I knew if I looked away, I wouldn't get another chance to see an event like this again, perhaps ever.

Being part of a large museum also gives you the opportunity to participate in other facets the museum has to offer. The Powerhouse Museum, of which Sydney Observatory was a part, had a train they sent all over the state, rarely coming back to Sydney. It had two carriages outfitted with an exhibition, one carriage with sleeping and kitchen facilities and a fourth carriage at the rear for storage and a bathroom. The train gave people in rural towns the chance to experience a museum like The Powerhouse without having to go to Sydney. It was a great success, and it was a shame when the museum closed it down. But while it was still operational, I put my name down for the occasional tour. After a couple of weeks of working on it every day, I would then come home, have four days off before going back to my regular job at the observatory. I loved being back under the wide-open skies of rural Australia, but the most vivid memory I have of those trips was of a time when the usually smooth operation of the train had a glitch.

Because the train would go to one town, sit there for a few days and then move on to the next town it wasn't cost-effective for the museum to have a locomotive attached all the time. So the carriages sat by themselves and when it was time to move an engine would come, hook up and transport the show to the next railway station or siding. This usually worked well. We knew when the locomotive would be coming and when it would get to the next town. For safety reasons, staff were not allowed to ride the train when it moved, so that meant we also had a car, allowing us to get around a town and drive to the next destination. We usually packed things up the night before moving, to protect them from being damaged during the trip, but since it only took half an hour to store everything away and this particular time the locomotive wasn't due until 10 am, we figured we had plenty of time to pack in the morning. But at 6 am we were awakened suddenly by a massive jolt caused by the locomotive attaching itself to the carriages. In the few minutes it took us to wake up and realise what was going on, the train started to move. All we could think of was we weren't ready. Besides not being dressed, we hadn't stowed things away and we were about to leave the car 200 kilometres behind at the station. In a matter of seconds, my boss decided he would get off and grab the car while I stayed on the train and secured everything. I don't think anyone has ever dressed so quickly, grab the car keys and jumped off a moving train. I was impressed. Left on the train, I had time to leisurely get dressed and go through the carriages, securing them before the trip got too rough. It was unplanned but a great journey between the two towns on the train that day. Country train rides are a fabulous way to relax and enjoy the countryside.

The night sky is generally predictable. The stars, planets and moon all move with consistent and comforting certainty about the sky. However, occasionally the heavens can offer up something that takes the world by surprise. In the next chapter, we will talk about one such event.

CHAPTER 5

Comet Shoemaker-Levy 9

In mid 1994, an event occurred that took the entire world by surprise. A comet was discovered that had passed a little too close to Jupiter and was torn apart by its massive gravitational field. It had produced a string of cometary fragments that had been calculated to crash into Jupiter on its next pass. No one knew what to expect. Would the impacts be visible, or would we see nothing? Would it be a public relations nightmare or boon? The difficulty with any new astronomical event is it is difficult to predict what, if anything, will be visible. Comets, in particular, are notoriously fickle for this. What would be seen when one crashed into Jupiter was anybody's guess.

The difficulty we had at Sydney Observatory, therefore, was just how much we invested in both time and advertising in the event. Given the one-off nature of the experience, we eventually decided to go all out and hold a major event where people could come and see Jupiter on each of the predicted six nights of impact. It was the correct decision, as things turned out spectacularly well. The weather co-operated brilliantly, all six nights were crystal clear, and Jupiter was in the perfect position to look at all evening.

Given the publicity the comet impact was generating we were expecting maybe a few hundred visitors per evening. And since we had advertised we would only be open for four hours each night, we felt we needed everyone working in order to get whoever showed up through in that time. Altogether there were 32 full-time, part-time and casual staff working each evening.

To keep everyone who came along occupied, all the staff had specific jobs. Usually, we collected entry fees inside the building, but for this event, we put a table down at the front gate so we didn't clog things up with people trying to pay inside. That required two staff. We had talks running continually in the theatre. That occupied two more staff, who alternated between giving the talk and crowd control for the people waiting outside the theatre. Three more staff wandered around the exhibition space, helping wherever they could. Five people were on general crowd control and answered questions outside with the telescopes. Two people did nothing but circulate between the different stations, running errands, delivering messages and providing relief for staff that needed to go to the bathroom. One person was in charge, making decisions and securing money collected from entry fees. But perhaps the best decision we made was to have two people assigned to do nothing specific, just floating around doing anything they saw needed to be done and alerting the appropriate staff if they saw potential problems arising. Everyone else operated the two large telescopes in each of the domes as well as five smaller telescopes in the yard of the observatory. If it sounds like it was busy, it was.

On the first night, we were ready to open at twilight. Already we could see there was a queue outside the gate, but we didn't appreciate just how long it was. As the night set in, we realised our estimate of a few hundred people per night was a dramatic underestimate. There was always a queue out the gate, at times about 400 metres long stretching to the bottom of the hill. Four hours was never going to be enough for the number we had. By the time we closed that first night, we had counted over 1,000 people had come through the gates, and it was 2 am in the morning.

I lived about 40 minutes away, so by the time I got home and crashed into bed, it was about 3 am in the morning. By itself, this wouldn't have been a problem except that I had to be back at work by 10 am, which meant leaving home at 9 am. Five hours of sleep was barely enough to recover.

The first night was a portent of what was to come. Every night we would start at 6 pm, at twilight, and finish around 2 am. Every day I would

Standing beside my telescope showing visitors the Comet Shoemaker-Levy 9 impacts on Jupiter.

have to be back at work by 10 am and do it all again. By the end of the six nights, I was so tired I was working on autopilot. It was exhausting, but one of the best working weeks I have ever had.

All up, over the six nights, over 6,000 people visited the observatory to watch Comet Shoemaker-Levy 9 crash into Jupiter. Amazingly there were only minor issues, and I put that down to the organisation and resources we had put into every evening. As a bonus, the entry fees taken were enough for the observatory to buy a new telescope and some other equipment we needed.

The view through the telescopes was spectacular and we found the best view was at the very beginning of the evening, during twilight. At that time the seeing was a lot steadier, and because the sky was still a bit light there was less contrast between Jupiter and the background, so it made the detail on the planet easier to see. The impacts themselves occurred just out of sight around the edge of Jupiter. Fortunately, it was on the side that soon rotated into view, so we didn't have to wait too long to see the aftermath. On the first night, we anxiously watched to see if anything would be visible and lo and behold a black spot about the size of the Earth rotated into view. Awesome! Each night as we set up we excitedly looked to see what had happened in the last 16 hours and we were never disappointed. A new spot, or two, would be visible and by the end of the week, there was a line of spots circling Jupiter. These then hung around for a few weeks.

But as much fun as this was, I started to feel the urge to again do something for myself.

CHAPTER 6

Expanding horizons

The Central Australian experience had stirred up an appetite to be my own boss again and expand my astronomical education horizons. I loved working at Sydney Observatory, but I could also see a niche it didn't address and hence an opportunity to continue working there while also branching out privately. This involved something Sydney did not have, a planetarium.

While at the Ayers Rock Resort, we often thought we should have had a small planetarium to use on cloudy nights. Back in Sydney, where the weather isn't as good as The Centre, I wanted to do something privately that was astronomical yet weather independent, and a portable planetarium fit the criteria perfectly. A full scale, permanently located planetarium was out of the question. The cost of construction alone was, at the time, astronomical (pun intended!). Other people and organisations had tried to raise funds to build one in Sydney, but it was just too expensive. A smaller, portable planetarium though was within financial reach. Being portable meant I could travel to different locations, providing they had a space large enough to accommodate the five metre diameter dome and schools, libraries, shopping centres all jumped at the chance.

I operated the planetarium with my old business partner from Central Australia and between us we had great fun going to different places and

Our portable planetarium set up in a school library.

conducting planetarium sessions. The biggest problem we had was trying to fit these occasional days into our usual working hours, especially once we started to get more and more requests for us to visit. The other problem we had was having to talk almost non-stop all day. If we visited a school, it usually meant doing six sessions in the day so my voice never really got a rest. There were a few days when after a couple of sessions, my voice gave way, and I had to nurse it through the rest of the sessions, barely making it each time.

One day I was at a school with the planetarium when I had one of my more surprising interactions with young children. At about 190 centimetres tall and solidly built I'm a large man, and for this particular session, I was going to have a group of Year 2 students. I was inside preparing the planetarium for the group when I heard them assembling outside. Since it was a portable planetarium, made out of material and held up by air pressure provided by a fan, we couldn't have a door to get in and out. We had a tunnel instead. The tunnel was also held up by air pressure, so we didn't have to crawl, we could walk through the tunnel, albeit bent over. When I heard the group outside, I started along the tunnel to meet them and begin the session. The only problem was it never occurred to me before just how small a seven year old child was. As I got to the entrance of the tunnel and stepped out I straightened up from my hunched over stance to my full height, and immediately one of the children screamed in terror 'IT'S A GIANT!!!' and promptly started to cry and hide behind the teacher. It took the teacher about 10 minutes to calm the child down enough to go inside the planetarium. While that was happening, I decided it was better if I went back inside so when the group saw me again, I was sitting down in the semi-darkness. I will never forget that moment, and ever since then I have been very conscious that, to a small child at least, I can look intimidating… even if I'm not really a giant.

Another memorable moment working for myself in Sydney came in 2001 when my business partner and I provided a stargazing experience for an exclusive business party on an island in the middle of Sydney Harbour. Fort Denison, a famous island situated not far from the Sydney Opera House and the Sydney Harbour Bridge, commands perhaps the best view of the city from a prime location in the centre of the harbour. During the day it is open to the public, and I had been there a few times before, but this time we would be there at night with a private function.

The first challenge was how to get a large, heavy telescope out to an island. We couldn't use a commercial ferry, so we had to hire a water taxi. That way we could take our time getting the gear on and off the boat, plus we could go out and come back when it suited us. The organisers of the event didn't even blink at the extra cost, and when we got to the island, we knew why. This event was huge. They had imported a full-sized snooker table, an enormous bar and numerous other bits and pieces for entertainment. Our water taxi and overall fee were insignificant to the amount of money being dumped on Fort Denison that night. In hindsight, we regretted not charging more!

Given that we needed an unobstructed view of the sky, we commanded the prime spot on the island, the grassy area on top of the old fort. For the next five hours, we felt distinctly underdressed but well and truly welcomed and appreciated. If you ever get a chance to spend an evening on Fort Denison in the middle of Sydney Harbour do it. It is a magical experience.

Never let it be said my business partner and I were lazy. As if what we were already doing wasn't enough, we took on yet another project, running the Macquarie University Observatory. The university had a few small domes on campus that they used for students in their astronomy courses. They wanted to open them to the public, and because we only had work and no social life at this point, we opened the observatory every Friday and Saturday night for a couple of hours for people to come along. We enjoyed our time there, but it was also clear that we had finally overstretched ourselves. We stuck it out for about a year, but then it was time to move on.

Not long after we stopped with Macquarie University I felt it was time to move out of Sydney once again, this time for good. So in 2003 I ended up in a country town called Orange, located in the Central West of New South Wales. Unfortunately, old habits die hard, so not long after arriving in my new town I took on several astronomical ventures.

Domes of the Macquarie University Observatory.

CHAPTER 7
A move to Orange

I started by running some adult education courses at the local community college, as they didn't require much preparation or effort and I figured it was an easy way to begin getting involved in the community. Early on, I also became involved with a local school and the people I met there changed things dramatically. So much so that after only a couple of years I had to give up doing the courses.

Being involved with the school allowed me to conduct regular talks and viewings with some of the classes, something I still do. And once I got to know the teachers and they got to know me, it wasn't long before one of them introduced me to a parent. He was a geologist by profession but also a keen amateur astronomer. One day we sat down and talked about all things astronomical and during the conversation, he expressed his desire to start an astronomical society in the town. Since he was keen to do the administrative work, I said I'd help with the technical side of things. A single ad in the local newspaper and one month later, we had our first meeting.

The Orange Astronomical Society met every month and did the usual society things, but the beauty of being in a country town is that everybody knows everyone else. That means if you have an idea, someone in the society probably already knows who you need to contact to realise your idea.

During some of the meetings, a few members wanted to get carried away and build a large public observatory. From my experience, I knew an observatory on the scale they were imagining would cost a small fortune and not be financially viable, so I suggested a better alternative would be a planetarium. It could run regardless of the weather conditions, and the cost of planetariums had decreased dramatically over the previous 10 years, so it was now a feasible option. It was also something New South Wales lacked, at least on the scale and quality I was suggesting. There were a few smaller ones around, but none were fully digital. In all of Australia, the only ones at the time comparable to what I proposed were located in Melbourne, Brisbane and Perth. As the society members became more and more enthusiastic for the planetarium idea, we became less and less focused on the society, until we realised we needed to temporarily stop and restart the society once we had a proper location. In other words, once we had the planetarium.

Our vision for a planetarium was to create an educational and tourist facility that would not only complement but also enhance the astronomical heritage of the Central West. To achieve this, we formed the not-for-profit organisation Orange Planetarium Incorporated (OPI). The project slowly grew in size until plans included an 11 metre diameter, 80 seat capacity planetarium, a multiple telescope observatory, a lecture/workshop room, function room, display areas, gift shop and snack bar. Unfortunately, as time went on, the project shrank and morphed and grew and shrank multiple times until we finally settled on a design.

From the very start, the aim was to foster community understanding of astronomy and the sciences while offering a major attraction for visitors and tourists to the region. To better understand what it is we proposed, let's momentarily take a step back and talk about the difference between a planetarium and a public observatory.

A planetarium is similar to a library or museum where the benefits and 'profits' are measured in terms of the educational, cultural enrichment, and economic value offered to a community. It is an indoor facility that uses optical and computer technology to project and simulate stars, planets and other celestial objects onto a domed ceiling. They can range from portable units to modest buildings, to grand architecturally designed facilities, such as the Hayden Planetarium in New York. Programs can operate day and night, independent of weather conditions and in the comfort of cinema-style seating. Initially, planetariums just used optical projectors, but today there are more innovative projection systems. In particular, digital systems that utilise the capabilities of modern computers. The original projectors were only able to project star fields, with additional images provided by slide and video projectors. Today, digital systems can show much more than just star fields and astronomical material, making it possible to present shows on a wide range of different subjects.

Early in the planning stages, we realised that to complement the planetarium the facility should also have a public observatory. A public observatory is any place that allows the public to view the sky through telescopes. It can be a large building with multiple domed rooms or an open area, giving a clear view of the heavens. It can also be located anywhere, irrespective of ambient light.

Of course, if you are going to have a planetarium, you need somewhere to put it and selecting a plot of land was not going to be straight forward. Eventually, after many years of discussions, negotiations and site evaluations, Orange City Council finally granted us a patch of land. Altogether over 17 sites were considered before everyone settled on the current option. One, in particular, was on the top of the local mountain. Someone thought it would be a great idea to have a tourist attraction at close to 1,400 metres. Given I live at 800 metres and it is heavily snowing as I write this, I don't think it would have worked. Trying to get people up a snowy mountain in the middle of winter was a big ask, especially given that the council closes the road whenever it snows.

The first serious option was at the Orange Botanic Gardens. It offered the best overall outcome in terms of location, ease of access, parking, prominence and complementary to existing facilities. Unfortunately, it turned out it was only the chosen site for a couple of years. We now have a new, final location located in the centre of town.

Now that we had the idea, knew roughly what we wanted and had somewhere to put it, the next thing we needed was a design. As a group of residents, we didn't have the funds to build a planetarium ourselves, so that meant trying to get funds from outside sources. But before we could ask for funds, we needed a design so we knew just how much funds were required. At this point, someone in the group had the brilliant idea of

One of the first site options offered as a potential location for the planetarium.

An artist impression of designs for the planetarium facility at the Orange Botanic Gardens.

contacting universities that had schools of architecture and asking them if they would like to create designs as part of a student project. Within minutes of sending the emails, we had a response from a university in Sydney saying they thought it was a great idea and would use it as an exercise for their Masters students.

After several visits to Orange and multiple iterations of the designs, we ended up with something I thought was quite good. Shortly afterwards, we had costings, and we were away.

We lobbied hard with local and national businesses, Orange City Council, and the state and federal governments. Everyone we approached agreed it was a great idea, and they fully supported the concept, but someone else had to be first to commit money to the project.

We applied for government grants, but because we weren't a local council, we had no chance. A government official even told us that though the grant guidelines didn't explicitly say we weren't eligible to apply, there was no point in doing so as they would ignore our application. This was a bit of a setback, but by then I was getting pretty good at the politics of dealing with governments and organisations.

We also realised that for the whole thing to work, the people of Orange and surrounding districts had to support the project. Without their backing, it was never going to happen. Consequently, we started to make me and the planetarium project known to local businesses. It was purely a public relations exercise, as I didn't ask them for any money, and almost unanimously they could see the economic benefit of the project to the community. Along with the local businesses, we also needed to make ourselves known to the general public, so I began presenting a regular segment on the local radio station and writing a weekly column for the local newspaper. I wrote the newspaper articles for close on nine years, running up a tally of almost 500 articles. This wasn't as easy as it might sound. Since the word limit for each article was not very much, I would write what I wanted to say and then edit it dramatically to the allowable size. Each article also had to be clear and concise and about relatively simple topics. Try coming up with something different to say every week for nine years that didn't involve complex or abstract concepts (like black holes) and you will see it is not easy to do.

The radio segments weren't any easier, although they were a lot more

An artist impression of the interior of the Botanic Gardens design.

fun. Every Monday morning, around 9:15 am, I would go to the studio after the announcer had finished his breakfast show and record a five minute segment. At the start, I let him know beforehand what I wanted to talk about, but as time went on and we became more 'professional' I would simply show up and give him the questions I wanted him to ask. Of course, it was a discussion and he would always ask other things as well, but the provided questions were at least a good start. We recorded the segment on Monday, but it was played on Tuesday morning. We pretended we did it live, and most people thought it was, but often we would forget when recording and say things like 'tomorrow night' (meaning Tuesday night) when we should have said 'tonight'. So when I say we became more professional, in reality, we would regularly stop because we had made a mistake, stuttered, said the wrong thing, or I didn't know the answer to a question. But as the professionals we wanted to be, we carried on and edited the mistakes out later so it appeared to be a seamless live discussion by two skilled individuals. The radio segment lasted for close on seven years, and I am still good friends with the announcer.

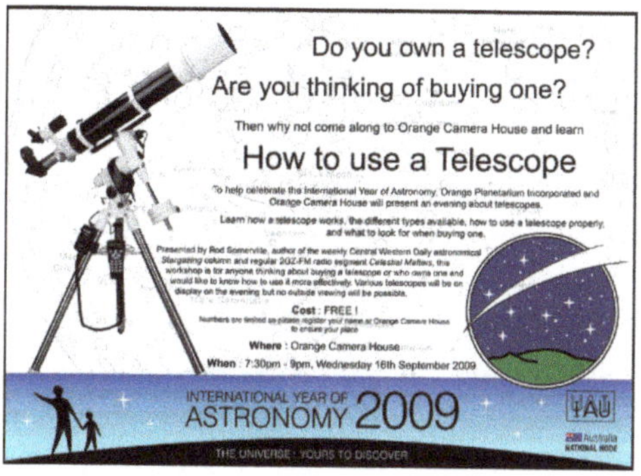

The International Year of Astronomy poster advertising our talk on telescopes.

Evening viewing during the Slow Summer festival.

Orange Planetarium Incorporated display at the Orange Rotary Expo.

In addition to the regular newspaper articles and radio segments, which were my area, our group also ran public activities to promote the project. We did events at schools and a public talk about telescopes as part of the International Year of Astronomy. One night we ran a public viewing night on the top of the Orange Regional Museum roof for no particular reason other than marketing. We had one day's worth of publicity and still managed to have around 200 people look through the three telescopes we had set up. Incidentally, the museum building didn't exist at the time we developed our Botanic Gardens design. When we finally saw the Museum designs we were shocked at how similar they were to our plans until we found out that one of the students who worked on ours was now a qualified architect and had worked on the museum designs.

The publicity was working. We were asked to give talks about the project to Rotary clubs, the local university as part of their lecture series, and to the Central West Astronomical Society for their annual Astrofest. Charles Sturt University also asked us to be part of an ongoing event by providing telescopes for an indigenous evening they run each year. But I think the one thing that gave me the most enjoyment was being asked to do some fundraising events for Ronald McDonald House.

At the time, Ronald McDonald House was in the process of fundraising for a new facility in Orange. The chairperson of the committee asked if I would be willing to help out. She said she wanted to reproduce here in Orange something she had done several years ago while visiting Uluru. While in the Centre she had attended a dinner in the middle of the sand dunes under the stars. At the end of the dinner, an astronomer gave a star talk and she loved it so much she wanted to recreate the same thing here. I couldn't let the opportunity pass so I asked what year she had been in the Centre and, lo and behold, it was when I was out there. It turned out my business partner had given the talk she attended. When I told her she was talking to the person who had started those dinner star talks she was super excited, so how could I not help out. She was even more excited when I said I would get my partner, who had given the talk she remembered from so long ago, to come from Sydney to help out.

The very first of these Ronald McDonald House events was held outside at one of the local wineries. The night was spectacular except for a slight problem. It was winter. When the sun went down the temperature plummeted to zero, probably less, and people were at risk of hypothermia. The tables were set for a formal dinner, but the decision was made to move everything inside. It was quite a sight to see the guests in their black-tie outfits pick up the tables and carry everything inside the warmer building.

Eventually, the publicity paid off as we finally convinced Orange City Council to back the project. The council then wanted to move it to a city block that already housed the local library, art gallery, museum and a theatre. We had no problems with this, as it would create a precinct that would be beneficial for all concerned. Not long after, the idea was raised that we join with the local Conservatorium of Music who were looking for a new home. Essentially we would be two separate entities under the one roof. This arrangement had many advantages, so we jumped at the chance. It meant the cost of the joint facility was higher than for just us, but joint facilities are apparently easier to fund. Of course, the new collaboration also meant new plans and designs, and I think the design we came up with is unique and we are more than happy with it.

Construction is about to start on the facility as I write this, so the next few years will be busy, but very exciting!

Over many years, I have learnt a lot from my diverse astronomical experiences. I would now like to share with you some facts, tips and ideas that you won't necessarily come across in your reading. Here are a few lessons I have learnt and some knowledge I have gained.

Artist impression of the exterior for the new
Orange Regional Conservatorium and Planetarium facility.

The site for the new Orange Regional Conservatorium and Planetarium facility.

Part Two
Lessons learnt, knowledge gained

CHAPTER 8

Curiosity is the key

One of the main things I have learnt over the years is that curiosity is the key. Find fascination in everything you come across and you will be well rewarded. There are no silly questions as every query, no matter how trivial or silly it might sound at first, can lead to a better understanding of the world around you. Quite often, they are also the right questions to ask.

I can still remember during the years of the Apollo program, when I was between the ages of eight and 12, watching the astronauts move about the surface of the moon and wondering why they were bouncing and moving so slowly. Why couldn't they simply walk about normally? It was a simple observation that sparked a thought that ultimately led to my discovery of gravity.

Similarly, a few years later I noticed something quite apparent that had eluded my attention up until that moment. We lived not too far from the beaches of Sydney, so during summer we frequently visited the ocean. I knew that the water in the Pacific Ocean was a deep blue colour but never gave it a thought why it was so until one day, while swimming in a friend's pool, the thought struck me that the water in the pool was clear. Why is water from the tap or in a swimming pool colourless yet the ocean was a beautiful blue colour? What was the difference? It took a few years of investigating, and several alternatives that ultimately turned out to be wrong, until I found the right reason while studying the physics of the atom and the nature of light at university.

For some of the questions I have had, I've yet to find an answer. For example, if the universe is expanding then the space between stars should also expand and the stars get further away from each other, but it turns out gravity overrides the expansion and keeps the stars together. So, given that gravity keeps the stars close but space is continually expanding (at an ever-increasing rate), is it possible to detect this motion of expanding space against the gravity bound stars? And another one. What causes inertia? I suspect the answer to this one may have come to light had I continued with my academic studies.

A slightly stranger question. If the horizon (about five kilometers distant out at sea) is the point where the earth's surface has curved so much that you can't see any more of it, why can't we see the curvature of the earth off to the side? In essence, every direction you look is about five kilometers away, so visually you have a large, flat circle of radius five kilometres surrounding you. Plus, the area within your vision is such a tiny part of a colossal sphere (about seven millionths the total surface area of the Earth) that it is impossible to notice any curvature it might have.

Some of the other thoughts and observations I have had over the years don't have quite so profound an answer. Some may never have an answer. The thing is, however, you don't know where the answer will take you unless you ask the question. Find fascination everywhere you look and the universe will teach you many wonderful things.

Another lesson I have learnt is that computer programs, apps for your tablet or phone, and computer-controlled telescopes are all useful and certainly have a place, but to get to know the night sky and genuinely appreciate it, you need to learn how to find your way around using just your eyes. You want to be able to go outside any night, look up and enjoy the heavens in all their glory. For me, some of the most pleasurable times have been just lying on my back and gazing at the stars.

But perhaps the biggest lesson is that it is okay to say you don't know. There are a lot of things in the universe to understand, so if you don't know, you don't know. Don't be afraid to say so. Over the years your bank of knowledge will grow, but you can't be expected to know everything, and most people don't expect you too. Whenever I get asked a question I have no immediate answer for I usually respond by saying 'That's a good question. I have a lot of information in my head, but unfortunately, that's not one of them. I will have to look it up'.

I've conducted lots of telescope viewings and come across a lot of people that claim I shouldn't be holding it here (wherever 'here' was at the time), and that I should be doing the tour out in the middle of nowhere because 'here' has too many lights. But I think they have missed the point. No matter where you are or how dark the sky, if it is clear you can always see something, even if it is just the Moon. Secondly, you do not need to be able to see faint fuzzy objects to get enjoyment from looking at the night sky. Sure, the darker the place, the better, but it is not essential. Even from a moderately large city such as Sydney, it is possible to have a great time looking at the stars. Since most people have never looked through a telescope before, anything they see is an experience.

If you are just beginning, or your main aim is to show a group of people the sky, then it is more important to be in a convenient location than it is to have a completely dark sky. If you make it as far away from lights as possible, it may very well be too far and you, or they, won't be bothered to travel the distance. If you are organising a viewing for a group, I have also found most people make up their minds whether to attend the evening at around 4 pm. At that time, if it is cloudy or raining it does not matter if it clears to be a perfect night by the allotted starting time, they have already decided it is not worth the risk and have settled in for the night. This used to frustrate me, but I now accept that is how it is and try to work out contingency plans just in case.

And remember the first law of using a telescope with a group of people: If there is one cloud in the sky, it will be over the object you want to look at, and if you decide to wait, it will never move. If you decide to look at something else, the cloud will move as quickly as possible to cover the new object. If you don't believe me, ask anyone who has regularly used a telescope!

Let's now look at some other information I have accumulated over the years. The topics discussed are by no means a complete coverage of everything in astronomy. That would be impossible. But hopefully, it is enough of a selection to get you started on your own list.

CHAPTER 9

Everything moves, always

One of the fascinating aspects of the universe is that it constantly moves. Nothing stands still and this motion affects us when we try to look at the sky. We can't simply go outside and expect things to be in the same place, every night, all night. If you want to make the most of your stargazing experience, you have to know what motions are in play and adjust accordingly. Understanding how everything changes and moves also teaches us a lot about the cosmos.

The first motion is probably the most obvious, the Earth rotates, resulting in the length of time known as a day. The Earth spins from West to East, which means from our viewpoint on the surface of the Earth, everything first appears on the eastern horizon and sets on the western horizon. That means everything rises in the east, moves across the sky and sets in the west.

But if it is us spinning and the stars staying still, why does it feel like we are the ones stationary and the stars move overhead? This idea is an entirely reasonable conclusion to reach, and it is easy to see how the Greeks and others assumed we were at the centre of the universe. It turns out if you are on the equator, due to the Earth's rotation you are moving at about 1,600 kilometres per hour, yet we cannot feel the motion. The shift to a sun-centred view of the solar system was a difficult one, as the universal experience contradicted the idea. It was more a case that shifting the centre to the Sun meant it was a more straightforward explanation of the motions of the planets. Ultimately this turned out to be correct, but definitive evidence for it took a long time to surface.

Because the Earth is rotating, everything in the sky appears to rotate about the point directly above the Earth's axis of rotation, that is, the point directly above the South Pole (or the North Pole if you live on the underside of our planet). Since very few people live at either of the poles, this point is not overhead but appears above the horizon the same number of degrees as your latitude. For example, if you live in Sydney at 33 degrees south latitude, the South Celestial Pole is 33 degrees above the southern horizon.

Now, the closer a star is to the celestial pole, the smaller the circle it makes around it. The further away you get, the larger the circle becomes until it looks like a star is moving in a line across the sky from east to west. So, the second motion we need to consider is that the stars are making circles in the sky with one complete circle taking one day. If you look at them through a telescope for more than a few seconds you have to keep moving the telescope in order to keep them in view.

The next motion is also related to the Earth spinning. According to our watches, one day is equal to 24 hours. This is true, but it is not as simple as you might think. If we look at the time the Earth takes to rotate once from a vantage point above the solar system it takes approximately 23 hours 56 minutes and 4 seconds. This four minute difference means, according to your watch, the stars rise four minutes earlier each night. If you were to go out at the same time each night for a year, you would notice the stars slowly change position, meaning as the months progressed, different stars and constellations would be visible. Eventually, after one year, you would be back to the same stars and constellations at the same time.

Even though I knew this intellectually and was aware that certain constellations were visible only at certain times of the year (such as Orion in the summer and Scorpius in the winter), it was beautifully demonstrated to me while working in Central Australia. Since I had to give tours of the night sky every night of the year, I quickly developed a spiel so I could move around the sky telling stories and pointing out stars in a way that flowed smoothly. The problem was, I couldn't use the same spiel for the entire year. As time progressed, constellations would become difficult to see then disappear while new ones came into view. Consequently, my narrative of the night sky had to evolve to accommodate the changing sky. Sometimes this wasn't easy to do and still maintain a smooth and consistent spiel.

As you watch the sky, you will notice that the planets move relative to the background stars. The planets, therefore, aren't always in the same position from night to night compared to the stars. In fact, the word planet comes from the ancient Greek term meaning 'wandering star'. This motion is caused by the planets being so much closer to us than the stars. While the star's motion from night to night is entirely due to the Earth's rotation, the planets are close enough that their movement around the Sun, combined with our movement around the Sun, causes the apparent position of the planets to change slowly.

The same thing happens with the Moon, although the effect is more pronounced due to its proximity to us. Look at the Moon for two consecutive nights and it is easy to see just how far it has moved compared to the stars. Since the Moon takes 27 days 7 hours 43 minutes and 41 seconds to do one lap around the Earth, which incidentally is the same time it takes to rotate once on its axis, each night the moon moves about 1/27 of the way around the sky. Divide 24 hours by 27, and you find that the Moon rises later each day by approximately 53 minutes. It does vary a bit depending on your location as the Moon's orbit isn't perfectly circular or in line with the Earth's equator, but you get the idea.

The final movement I want to mention concerns the Sun. The Earth is tilted by 23½ degrees compared to the plane of its orbit. Because of this, sometimes the southern hemisphere is tilted towards the Sun, and sometimes it is tilted away from the Sun. This is what creates the seasons. It also means the Earth's tilt is continually changing relative to the Sun. Looking from here on the Earth, this has the effect of varying the maximum height of the Sun above the horizon from day to day. This changing height also means that the Sun rises and sets in different locations on the horizon throughout the year.

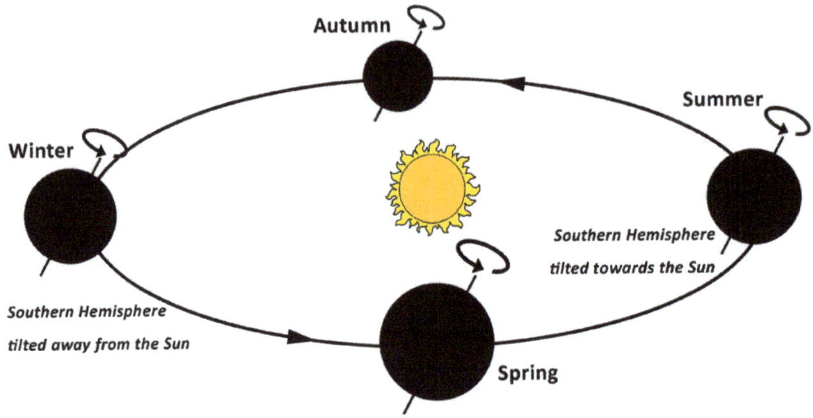

The tilt of the Earth's axis throughout a year.

CHAPTER 10

Astronomy, what is it?

Astronomy is one of the oldest sciences, if not the oldest, with references to astronomical events found in the myths, legends and beliefs of every civilisation. Essentially, astronomy was born when the first person looked up into the night sky and wondered what all those twinkling lights were.

Apart from some forays to objects within our solar system, astronomy is also the only science where we are not able to (currently) get out there and touch the things we are studying. We can only look. Also, in a lot of cases, we only know of one example. Despite this, it is remarkable just how much we have been able to find out. Merely looking and using our brains to analyse what we see has accomplished a lot.

The first astronomers used nothing more than the naked eye, and much was achieved in this way. However, the invention of the telescope enormously increased our understanding of the heavens. Another leap occurred with the invention of instruments that could then record and analyse the light coming through the telescopes.

Even though the realm of astronomy is as vast as the cosmos itself, it is a science where anybody with nothing more than a keen interest can still actively participate and contribute. Professional astronomers, with their expensive telescopes, cannot continuously scan the sky looking for an event on the off chance they will see it. Nor are they able to make repeated observations over an extended period of time. And sometimes the data they have is just too much for any one person or small group of people to ever possibly wade through. It is here that amateur astronomers with time, enthusiasm and sheer numbers can make a useful contribution.

And contrary to popular belief, it is not essential to have a telescope to discover the wonders of the sky. To see some things takes no more than your eyes, a dark sky and a bit of patience.

CHAPTER 11

What can be seen... easily?

Now that you are primed to look at the universe, what is up there for you to see? Some of the more exotic objects, such as black holes and neutron stars, do not appear in this list as they are beyond the reach of the average amateur.

Stars

Stars are the most numerous and obvious night time object. A star is a self-luminous ball of gas and they come in all sizes, temperatures, colours, brightnesses and distances. Stars are mainly made up of hydrogen and helium, but most have small amounts of other elements as well. About 6,000 individual stars are visible to the naked eye, although you can only see about 2,600 at any one time. Combinations of different distances, sizes and intrinsic intensity cause the varying brightnesses of the stars you see.

The Sun

The Sun is the closest star to the Earth and essential to us. It is the centre of the solar system and to date is the only star we have been able to study in any significant detail. Even though it is 1.4 million kilometres across and has a surface temperature of about 5,800 degrees, the Sun is a below-average star when compared to the other visible stars in the sky. The only real difference between the Sun and the other stars is simply one of distance. The Sun is very close, a mere 150 million kilometres away. The next closest star is about 40 million million kilometres away (4.2 light-years).

Planets

In our solar system, we know of eight planets, Earth being one of them. In recent years we have discovered thousands of planets around other stars, with the number of confirmed exoplanets rising almost daily. A planet is a small, relatively cool body in orbit around a star that only shines by reflecting starlight. In the case of those in our solar system, they shine by reflecting sunlight. Planets can be gaseous like Jupiter or solid like the Earth. I will also lump into this last category the Dwarf Planets such as Pluto, Ceres and Makemake.

Moons

A moon is a body in orbit around a planet. Most of the planets have moons around them, with only Mercury and Venus missing out. Some are quite substantial in size while the majority are about the size of a mountain or less. The easiest moon to see is, of course, our own, The Moon.

Meteors

Meteors are extraordinarily plentiful with millions of them occurring each year. They are also known as 'shooting stars' or 'falling stars', although these are bad names as they aren't stars. Most are nothing more than bits of dust or small rocks falling into the Earth's atmosphere and vaporising as they plunge through the air. Occasionally a larger rock falls into the atmosphere and may last long enough to hit the ground. These are then called meteorites.

Comets

Every so often, a comet becomes visible. These are large, icy objects in orbit about the Sun. Unlike the planets, however, comets have a very elliptical orbit, and it is this that makes the comet so spectacular. As the comet comes in close to the Sun, it heats up and its ices start to vaporise. This released gas and dust then gets blown away from the nucleus of the comet by the solar wind to create the spectacular tail that can sometimes be seen from the Earth.

Satellites

Thousands of artificial satellites now orbit the Earth. Many are visible from the ground at certain times and appear as a star-like point of light moving at a constant speed across the sky.

Open star clusters

Moving away from the solar system and heading into the bulk of the galaxy, we find that most stars cluster together. There are numerous groups of stars (usually holding in the order of a few hundred) visible all around the sky. Most of these groups have no particular shape to them, and because of this, they are called Open Star Clusters. The stars that make up these clusters are also very young. The most famous open cluster is The Pleiades, more commonly called 'The Seven Sisters', or in Japanese, 'Subaru'.

Globular star clusters

Globular Star Clusters are groups of stars that have a very regular shape to them. They cluster into a ball shape, often having up to a few million stars. They consist of very old stars, perhaps the oldest in the universe.

Nebulae

Nebulae are arguably the most spectacular objects in the sky to observe. They are large clouds of gas and dust and most people have at least seen pictures of one. Some produce their own light and consequently glow a red colour. Some scatter light from nearby stars and appear a beautiful blue colour (for reasons similar to why the sky is blue). Others don't have any light, either scattered or produced and appear black.

Galaxies

A galaxy is a massive collection of stars. In our galaxy, the Milky Way, our Sun is just one of about 400 billion stars. Galaxies are isolated objects that have vast distances of space between them and tend to come in only a small range of shapes: spiral, barred spiral, elliptical and irregular. To the unaided eye, there are only four galaxies visible, although some people claim to be able to see a fifth. These are the Milky Way (of course), the Large and Small Magellanic Clouds and the Andromeda Galaxy. If you are one of the people with super eagle eyes, the fifth is called M33.

CHAPTER 12

Generally speaking

—

The night is dark

On a clear evening as you look up, have you ever wondered why it is dark? Why isn't the night sky as bright as the surface of the Sun? In an infinite universe with an infinite number of stars, it should be.

Although individually stars appear feeble, it is only because they are a vast distance from us. But even though they look tiny and faint, the intensity of light we see from a star is the same as that emitted by the same sized area on the surface of the Sun. If you were to move the Sun twice as far away, we would intercept only one-quarter of the light, but it would also appear only one-quarter the size, so the intensity remains constant. With a distant star, the same applies.

So if the universe is infinitely large and has stars everywhere, then no matter what direction you look, you should eventually see a star and the entire heavens should be as bright as the Sun. Johannes Kepler realised this in 1610, but it became famous in 1826 when Heinrich Olbers took up the issue. Ever since then, it has been known as Olbers Paradox.

Many possible explanations have been offered. Olbers himself thought dust between the stars absorbed most of the distant starlight. But this can't be true as the dust itself would eventually heat up and emit light. Other explanations include the universe having only a finite number of stars, or the distribution of stars not being uniform, such as some hiding behind others. The problem with these explanations is that limited as they may be, or even if there is an infinite number hiding behind each other, there would still be enough stars to make the sky extremely bright.

The solution to Olbers Paradox, and the real reason the night sky is dark, involves a couple of other ideas. The first and main one is that the universe is not infinite. It is still relatively young, only 14 billion years old. Consequently, we can only see the stars that started emitting light at most 14 billion years ago. Due to the expansion of the universe, the actual distance to these stars is now more than 14 billion light-years away, but the argument still holds. The light from any stars lying further than this hasn't had time to reach us yet. Secondly, stars do not last forever, especially the brighter ones. They typically burn out after a few billion years, so some of them have already switched off.

These two reasons combined mean that once the sun goes down, we have a wonderfully dark night sky rather than having to live in perpetual light.

Latitudinal effects

Most people have the impression that the stars we see in Australia are entirely different from those seen in the northern hemisphere. However, just because you cross the equator doesn't mean one set of stars disappear and another set comes into view. So what exactly is the situation?

The Earth is surrounded by stars so if you were to stand on the equator, over one full day they would theoretically all be visible. In reality, all sorts of factors (such as topography, light pollution, atmospheric extinction, and so on) would prevent you from seeing them all.

As you then walked from the equator towards the south, you would start to lose the stars visible from overhead at the north pole, losing sight of more and more the further south you walked. For example, Sydney is 33 degrees south of the equator. So, if you lived in Sydney, you could see all the stars except for those within 33 degrees of the north celestial pole (the point directly over the north pole on Earth). You would be able, for example, to just see some of the stars in the pattern known as Ursa

Major, the Big Bear (also known as the Big Dipper). Similarly, people in the southern USA or Europe can see the Southern Cross. There are reasons why in practice you may not be able to, but in theory, you can.

As you continue heading south, you wouldn't gain any more stars as you were already able to see the stars directly overhead at the south pole when you were on the equator. Admittedly they were right on the horizon, but this is a theoretical exercise, not a practical one. By the time you arrive at the south pole, you would be able to see, at most, only half of all the possible stars in the sky, only those south of directly overhead at the equator. To see the other half you would have to be able to look through the Earth, and this is at best impossible to do. The same applies if you headed north, but with the other half of the stars.

An easy way of determining whether a star lies in the southern or northern part of the sky makes use of the group of stars known as The Saucepan (Orion's belt and sword). Since it lies almost directly over the equator, any stars south of it are over the southern hemisphere, any stars north of it are over the northern hemisphere. It is ever so slightly on the southern side of the celestial equator, but it is close enough to use as a guide.

So, if we can see most of the stars surrounding the Earth from most places on the Earth, why does the sky look different when we cross the equator? The answer to this involves a combination of things. Firstly, as you move further north, the stars that were overhead when you began now appear lower in the southern sky and some new stars have appeared, which can be quite disorientating. Also, when you cross the equator, you effectively turn upside down, and any star patterns are now upside down as well. Combine these two effects, and it is easy to start thinking the sky has completely changed.

Incidentally, when you do find Orion in the sky, you will notice that he is standing on his head. Orion turns out to be one of the few constellations that can be seen from everywhere on Earth, so the stars are easily visible from both the southern and northern hemispheres. They were therefore first made into a constellation by people in the north, condemning us here in the south to always see Orion hanging upside down. This is true for most of the constellations. It is only the handful created when sailors from the north came down to southern waters that we get to see any the right way up.

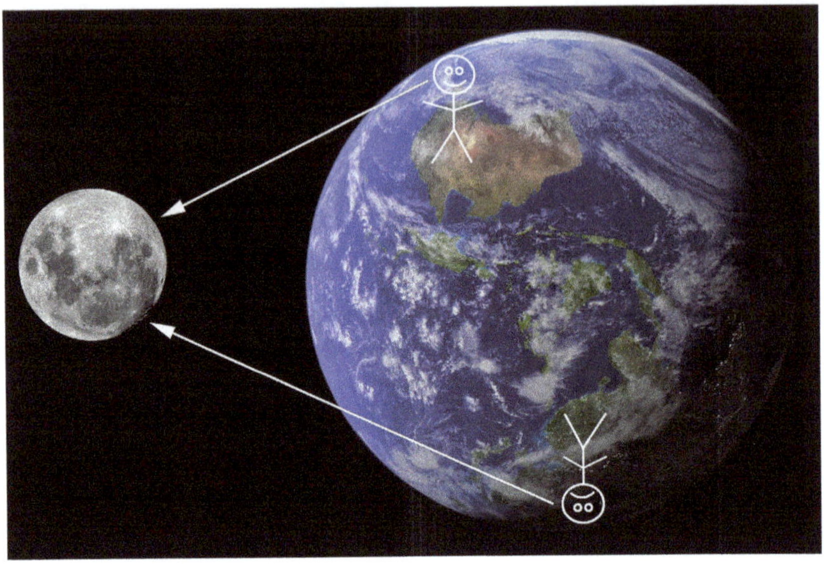

The moon and stars look upside down from the other hemisphere because they are the other way around.

Southern supremacy

Have you ever wondered who has the best night sky? Who has the most interesting things to look at, the southern or northern hemisphere? It may seem biased, but, of course, we do here in the southern hemisphere.

Larger and easier to see than any visible in the northern sky are two of the closest galaxies to our own, the Large Magellanic Cloud and the Small Magellanic Cloud. They appear like bits of the Milky Way that have broken off.

Still in the south, but a bit harder to locate, are the two best globular clusters of stars in the entire sky. Found near the Southern Cross, Omega Centauri and 47 Tucana are two of the most beautiful things you can view through a telescope.

Even when it comes to individual stars, we in the southern hemisphere hold the upper hand. The three brightest stars in the sky reside in the south with us. Sirius is the most brilliant and found near the group of stars known as The Saucepan (Orion's belt). Canopus, the second brightest, is located in the constellation of Carina. It is a genuinely magnificent star. Over 14,000 times brighter than the Sun, it is also about 65 times as wide. Which brings us to the third brightest and arguably most interesting star, Alpha Centauri. Easily located by most people, it is the pointer star furthest from the Southern Cross. As far as stars go, it is relatively average, except for a couple of details. Firstly, it is a triple star system in which the two brightest stars make the finest example of a binary star anywhere. But more importantly, it is our next-door neighbour, the closest star system to the Sun.

And one last reason why the south has a better night sky. Early evening during winter has the centre of the Milky Way galaxy in the best possible place to view it, directly overhead. Something the north never gets to experience.

How far is far?

Ask anyone how far they can see, and I guarantee that after a moment of thought almost everyone will quote something in the order of fewer than 50 kilometres, primarily thinking about their estimate of the distance to the horizon. But stop and think about this for a moment and you will see how much this answer reflects our earthbound prejudice.

Assuming normal vision, if we look up, instead of along the ground, we can, of course, all see the Moon. At roughly 384,000 kilometres, it is a little bit further than the horizon. If we look towards the Sun, our eyesight extends to approximately 150,000,000 kilometres (don't ever actually do this as it is far too dangerous, but hopefully you can see what I'm getting at). At roughly 1,500,000,000 kilometres the furthest planet we can see easily with our eyes is Saturn. Continuing out of the solar system, at a mere 40,000,000,000,000 kilometres distant, we can see the closest star system, Alpha Centauri. All the other stars we see in the night sky are even further away.

Already we are well past the initial estimate of the horizon, so what is the furthest thing we can see with our eyes? The answer is the Andromeda Galaxy. Lying roughly 24,000,000,000,000,000,000 kilometres away, it is definitely further than our Earthbound bias first suggested. Some people say they can see another galaxy slightly further away called M33, but they are the exception.

Quoting all these distances in kilometres eventually becomes quite cumbersome. I know it was for me when typing them out. The Andromeda Galaxy is also by no means the furthest object in the universe. To make life easier, astronomers invented another way of measuring distances in space, the light-year. One light-year is the distance light travels in one year and is equal to 9,460,730,472,580.8 kilometres. The distance to the Andromeda Galaxy now becomes a much more manageable 2½ million light-years.

Another way of quoting distances in space is to talk about how long it takes light to get there. Using this approach, the Moon is only 1¼ light seconds away, the Sun eight light minutes, Saturn 84 light minutes and Alpha Centauri 4.3 light-years distant.

No matter what metric you use, astronomical distances are challenging to comprehend. To put them into perspective, let's use the distance between the Earth and Sun as our basic unit. If this distance were equal to one metre, the Moon would be only three millimetres away, Saturn just 9½ metres away, Alpha Centauri would be about 270 kilometres, and the Andromeda Galaxy would be on the Sun. On this same scale, the Earth would be 1/10 of a millimetre across and the Sun a mere nine millimetres.

Makes you appreciate how big space is, doesn't it?

What's in a name

How do we name all the things we find in the sky? This may sound trivial, but it is quite important. In case you hadn't realised, there are an awful lot of things up there and if you try to find something celestial in a book or online, you very quickly run across a confusing jumble of names, numbers and combinations of both. It isn't necessary to know the names of all the objects, but it is crucial to recognise at least what someone might be referring to when they throw an obscure name or number at you.

All the names of the five naked eye visible planets, the Sun and the Moon come from Roman and Greek gods. As new planets were discovered, this naming system continued. Most of the moons are also named after people or creatures closely associated with the god they happen to be circling. Unfortunately, there are a few exceptions. For some reason, the names of the moons around Uranus were given names (by John Herschel in 1852) from works by William Shakespeare and Alexander Pope. When new Moons were discovered by the spacecraft Voyager 2 in 1986, the International Astronomical Union (IAU) decided to continue this system.

With a newly found asteroid, the discoverer gets to nominate a name for it. The name has to be then agreed upon by the IAU before it becomes official. You are not allowed to name an asteroid after yourself or political and military people. Initially, most of the asteroids also followed the mythological naming system, but now names include prominent scientists, famous people and even places.

The only things in space that can be named after you are comets. If you discover a comet and are the first person to let the rest of the world know about it, then it is named after you (e.g. Comet Encke). Even spacecraft can have comets named after them (e.g. Comet SOHO). If more than one person discovers it at the same time, both names are associated with it (e.g. Comet Hale-Bopp). Up to three people can have their names given to a comet (e.g. Comet Tuttle-Giacobini-Kresak). If you or perhaps a pair of people working as a team discover more than one comet, then it is your name(s) and the running tally of how many you have found (e.g. Comet Machholz 2, Comet Shoemaker-Levy 9).

Some objects have common names, as you might expect (e.g. the Andromeda Galaxy, the Swan Nebula and the Jewel Box). But there are an awful lot of objects, so it isn't convenient to give them all common names. An alternative is to assign them a catalogue number.

In 1780 a French amateur astronomer, Charles Messier, was looking for comets. At the time, it was what all people of noble birth were doing. But the method by which you find a comet is to look for a fuzzy object. If you find one, you then observe it over a few nights and see if it moves compared to the background stars. If it does, chances are you have a comet. If it doesn't, well, tough luck, it's something else. Messier kept finding a lot of fuzzy objects that would turn out to not be comets, so he made a catalogue of them to tell people not to bother looking. His catalogue ended up with 109 entries. The first was called M1 (the Crab nebula), the 22nd was called M22, and so on. These 109 objects are much more interesting than the comets he was after since they turn out to be galaxies, nebulae and star clusters. Some other well-known examples are M31 (the Andromeda Galaxy) and M42 (the Orion nebula).

In 1888, another catalogue was created. This time it listed approximately

8,000 objects, including all of the Messier objects. It was called the New General Catalogue (not so new now of course). The objects, therefore, had an NGC number. Some examples using this naming system are NGC1952 (the Crab nebula), NGC224 (the Andromeda Galaxy) and NGC1976 (the Orion nebula). Some other famous, but distinctly southern hemisphere examples are NGC5139 (Omega Centauri), NGC4755 (the Jewel Box) and NGC5128 (Centaurus A galaxy).

In about 1900 an addendum was made to the New General Catalogue, creating the Index Catalogue. Objects listed in this have an IC number. Another, later, catalogue was the Barnard Catalogue. These had B numbers.

Things have become messy because as each catalogue was made, the previous ones were never abandoned. So any particular object may have dozens of names and catalogue numbers. Fortunately, the main ones used and the ones most people will ever come across are the Messier and NGC catalogues.

Some stars have common names. These are mainly the brighter ones and are limited to a couple of thousand at most (e.g. Sirius, Canopus, Arcturus, Vega, Capella, Rigel, Betelgeuse). Some are named after the people that discovered them (e.g. Barnard's star, Kapteyn's star, Wolf's star). But once again, with about 6,000 individual stars visible to the naked eye alone, there are just far too many to have a common name. So how do we name them?

One way is to use the constellation in which they lie. If you call the brightest star in the constellation 'alpha' and then a derivative of the constellation name, you have a naming system. The next brightest would be 'beta' and so on (e.g. Alpha Centauri, Gamma Crucis, Epsilon Eridani). The obvious limitation with this method is that there are only 24 letters in the Greek alphabet.

One way of getting around this problem once again uses the constellations, but this time, starting from one side and moving a line across the constellation in increasing Right Ascension, number the stars as you bump into them. So the first one would be '1' then the derivative of the constellation name. The second would be '2', the third '3' and so on (e.g. 61 Cygni, 40 Eridani, 47 Tucana). Using this method gives an unlimited number of names, but what happens if you get through naming them all and some conscientious astronomer finds another star back in the middle? Do you rename all of them? Do you tell the astronomer to forget they ever found it? No.

Today there exists a slightly different system. All stars are now basically located and listed by their coordinate in the sky. Since this method doesn't rely on any specific order for the naming process and is virtually unlimited, it is possible to give a name to every star found. It may be a practical way of naming things, but I think its most significant problem is that the names are just, well, ugly. There's no other way to describe them. For example: Sanduleak -69 202, BD+38° 3238, and HD 45348. This last star is also known as Canopus, the second brightest star in the night sky. HD 45348 or Canopus? I know which name I prefer.

Celestial Coordinates

Since everything in space is so far away that our eyes cannot tell distance, all the stars and planets look like they are on the inside of a giant celestial sphere surrounding the Earth. However, in reality, they are scattered in distance as well as direction. But when the only thing you had was your eyes and no means of calculating distances to the stars, a celestial sphere was a reasonable model of the cosmos and it is still a useful concept to use. Due to this idea of a celestial sphere, in the same way the Earth has a grid system of latitude and longitude, the sky also has a grid system, known as Declination (Dec) and Right Ascension (RA).

Declination is simply an extension of the Earth's latitude lines onto the imaginary celestial sphere. Directly above the equator on the Earth (0 degrees latitude) is the celestial equator (0 degrees Declination). Directly above the South Pole (-90 degrees latitude) is the South Celestial Pole (-90 degrees Declination). At any point in between, directly overhead would be the same in Declination as your location is in latitude. For example, directly above Sydney (-34 degrees latitude) is -34 degrees Declination.

Right Ascension is a bit trickier. You cannot merely extend the Earth's lines of longitude since the Earth is continuously rotating. That would have the lines of Right Ascension continually sweeping across the sky. The solution is to pick a point that a line of Right Ascension passes through going between the south and north celestial poles and call it zero. Once the point is selected, you can then go around the sky drawing your Right

Ascension grid. Essentially this is what we did on Earth for zero degrees longitude, which passes through Greenwich in England.

The zero Right Ascension point chosen in the sky is called the First Point in Aries. It is one of the points where the imaginary celestial equator intersects with the projected line of the plane of the planets, the ecliptic. There are of course two intersection points, but the one chosen was the one initially in the constellation of Aries. It has, however, since moved into Pisces. Why has this point moved? Since defining the point as the intersection of two lines, if either of those lines move, then the intersection point will also move. The line of the ecliptic is not likely to change as that would require all the planets to rearrange their orbits. Therefore, it has to be the celestial equator that changes.

As the Earth spins, its axis of rotation wobbles slowly, similar to the axis of a spinning top as it slows down. This wobbling is relatively slow, taking approximately 26,000 years to do one wobble, so generally we don't notice it. But one of the consequences of this movement is that the Earth's equator, and by extension the celestial equator, slowly changes its orientation in space. This slow change means the point used as the zero for our celestial coordinate system also gradually changes and is known as the precession of the equinoxes.

Making one complete wobble every 26,000 years is a long time for a human, so we don't notice things changing on a day-to-day basis. But for astronomers with extremely accurate tracking mechanisms on large telescopes this ever so slight change is noticeable. Consequently, the coordinate system used to locate stars and objects has to be updated every 50 years to compensate for this slow drift.

When it was first worked out back in about 160 BCE, the First Point in Aries used to reside within the constellation Aries, hence its name. In 68 BCE however, it moved to be within the boundaries of Pisces, where it will remain until the year 2597 CE when it moves into Aquarius. Since the year 2597 CE is not quite 600 years away, but the year 68 BCE was almost 2,100 years ago it means we are getting reasonably close (let's not nitpick) to the time this point moves from the constellation Pisces into Aquarius. As a piece of trivia, this is the origin of the astrological phrase 'Dawning of the Age of Aquarius'.

CHAPTER 13
Earthly matters

Blue sky

It is easy to overlook something we live with all our lives. For example, have you ever wondered why the sky is blue?

Many attempts to explain our blue sky have been put forward, but the real explanation has to do with the Earth's atmosphere itself. Light coming from the Sun is composed of all different colours, which we get to see whenever we look at a rainbow, or when sunlight passes through a prism. As sunlight travels through the atmosphere, it hits a lot of small particles which cause the different colours to scatter by different amounts. Blue light gets scattered more than the other colours, so when we look up during the day, we see more blue light.

But what about sunsets? When we look at the horizon at sunset we are looking through a lot more of the atmosphere and the light has had to travel further through the air than it does in the middle of the day. By the time it gets to our eyes, the extra air has done an excellent job of scattering away all of the blue light. However, red light, which is not scattered nearly as much, can still penetrate the atmospheric haze, making sunsets look red. The same also applies to sunrises.

If the small particles that scatter light are sufficiently abundant, there will be almost the same amount of blue and red light scattered, making it appear white. This is why clouds are typically white.

Incidentally, there is no distinct boundary between the atmosphere and space. The atmosphere just progressively gets thinner and thinner until eventually you think to yourself 'Hey, I'm in space!'. However, even at their great heights, satellites and the International Space Station still experience some drag due to an extremely thin atmosphere. So, since the atmosphere just slowly thins out rather than abruptly stopping, the edge of the atmosphere has been defined as occurring at an altitude of 100 kilometres above the Earth's surface.

Also, the density of the atmosphere does not drop off uniformly as you go higher. Even though the boundary with space is 100 kilometres (100,000 metres), it turns out that 50 per cent of the atmosphere lies below 5,400 metres and 90 per cent below 16,500 metres. To give some reference to these numbers, the summit of Mount Everest is 8,848 metres (and above 58 per cent of the atmosphere), commercial planes typically cruise at about 10,000 metres (above 72 per cent), and the telescopes on the top of Mauna Kea in Hawaii are at 4,200 metres (40 per cent). So, if you want to be a mountaineer or a professional astronomer, you have to be prepared to breathe some thin air.

One day

Most of us are familiar with the concept of a day, but what is its origin? One day is defined as the time it takes the Earth to rotate once on its axis. The problem is, however, that measuring this period is not as easy as you might expect.

If we use the stars to measure the time it takes the Earth to rotate once, we end up with a period known as a Sidereal Day of 23 hours 56 minutes and 4 seconds long, and this reflects the actual rotation of the Earth. Humans, however, are primarily daytime creatures. If we measure the rotation period using the Sun, we find that things aren't as straightforward.

Tracking from day to day when the Sun is at its highest point in the sky shows that the length of the interval between successive highest points varies quite a bit. This variation is mainly due to the Earth's changing speed as it goes around the Sun. If we take the average time between successive daily transits by the Sun, we find it is equal to the familiar 24 hours and is called a Mean Solar Day.

The differences between the mean solar day and the sidereal day is also due to the Earth's motion around the Sun. In the time the Earth has rotated once it has also moved around its orbit a little bit. For the Sun to be overhead again, the Earth has to, therefore, rotate once plus a little bit more. If you divide the length of a sidereal day by 365.256 days (the time it takes for the Earth to do one lap around the Sun), you get 3 minutes and 56 seconds. Add this difference to a sidereal day, and you end up with the familiar 24 hours.

The Earth's rotation, however, is slowly slowing down. Caused by the ocean tides, it is continually decelerating. Modern clocks are now a much more accurate way to keep time than using the Earth's rotation and what we find is that after about 500 days the difference between the Earth rotation time and the atomic clock time is about one second. Instead of allowing this to accumulate, a leap second is inserted to bring the two times closer together. This leap second can be either plus or minus, although, unless someone told you it had been introduced you would never notice.

Days of the week

Thousands of years ago there were seven known 'planets' that moved around the sky: the Sun, the Moon and the five planets visible to the naked eye. With seven days in a week, there must surely be a connection of one day for each of these planets. So, believe it or not, every time we mention the names of the days of the week, in a roundabout way we are talking about the planets.

There is no way to mistake which planet Sunday is named after. Only a little less obvious is Monday, named after the Moon. Saturday named for Saturn isn't too hard to pick either. After that, things get a bit tougher.

The other four days don't sound at all like the remaining planets and that is because they get their names from Norse gods. When northern European countries adopted the seven day week, they substituted what seemed like the best Norse equivalents of the Roman gods. Tiw, the Norse god associated with war, replaced Mars and gave us Tiw's day, or Tuesday.

Odin (Woden) became associated with Mercury so that Mercury's day became Woden's day (Wednesday). Thor was identified with Jupiter and made Thor's day (Thursday). Finally, Friday comes from Frig's day, for Frig, the goddess of fertility (in a lot of ways similar to Venus).

How the order of the days within the week came about is even more convoluted. Assuming everything rotates around the Earth (since that's what it looks like), simple observation shows the planets take different amounts of time to move around the sky. That means, if we want to, we can list them in order of decreasing times they appear to take to move around us. Taking the longest to move around the sky means Saturn comes first, then Jupiter, Mars, the Sun, Venus, Mercury and finally the Moon.

Originally the planets were associated with each hour of the day, and it was customary to name the whole day after the planet of the first hour. If the first hour of the day was associated with Jupiter, then the entire day was called Woden's Day (Wednesday). So, if we start with Saturn (because it takes the longest) for the first hour of the first day, we have Saturday. Working through the hours and the order means we have Mars for the 24th hour. If we continue through the order, we find the first hour of the next day is associated with the Sun, making it the Sun's Day, or Sunday. Continuing this process for seven days, we find the familiar order of the days of the week soon become apparent.

Since I know you want to do it, to save you the trouble of writing it out, Table 1 has the entire week hour by hour.

Hour of day	Day 1	Day 1	Day 1	Day 1	Day 1	Day 1	Day 1
1	Saturn	Sun	Moon	Mars	Mercury	Jupiter	Venus
2	Jupiter	Venus	Saturn	Sun	Moon	Mars	Mercury
3	Mars	Mercury	Jupiter	Venus	Saturn	Sun	Moon
4	Sun	Moon	Mars	Mercury	Jupiter	Venus	Saturn
5	Venus	Saturn	Sun	Moon	Mars	Mercury	Jupiter
6	Mercury	Jupiter	Venus	Saturn	Sun	Moon	Mars
7	Moon	Mars	Mercury	Jupiter	Venus	Saturn	Sun
8	Saturn	Sun	Moon	Mars	Mercury	Jupiter	Venus
9	Jupiter	Venus	Saturn	Sun	Moon	Mars	Mercury
10	Mars	Mercury	Jupiter	Venus	Saturn	Sun	Moon
11	Sun	Moon	Mars	Mercury	Jupiter	Venus	Saturn
12	Venus	Saturn	Sun	Moon	Mars	Mercury	Jupiter
13	Mercury	Jupiter	Venus	Saturn	Sun	Moon	Mars
14	Moon	Mars	Mercury	Jupiter	Venus	Saturn	Sun
15	Saturn	Sun	Moon	Mars	Mercury	Jupiter	Venus
16	Jupiter	Venus	Saturn	Sun	Moon	Mars	Mercury
17	Mars	Mercury	Jupiter	Venus	Saturn	Sun	Moon
18	Sun	Moon	Mars	Mercury	Jupiter	Venus	Saturn
19	Venus	Saturn	Sun	Moon	Mars	Mercury	Jupiter
20	Mercury	Jupiter	Venus	Saturn	Sun	Moon	Mars
21	Moon	Mars	Mercury	Jupiter	Venus	Saturn	Sun
22	Saturn	Sun	Moon	Mars	Mercury	Jupiter	Venus
23	Jupiter	Venus	Saturn	Sun	Moon	Mars	Mercury
24	Mars	Mercury	Jupiter	Venus	Saturn	Sun	Moon
Name:	**Saturday**	**Sunday**	**Monday**	**Tuesday**	**Wednesday**	**Thursday**	**Friday**

Table 1: Days of the Week

Twilight

Wherever you live, once the sun goes down, it takes a little while for the sky to get dark and the stars to appear. If you visit somewhere close to the equator, however, it is amazing how much quicker it gets dark compared to somewhere further south of the equator. Why the difference?

Firstly, let's talk about how we work out sunrise and sunset times. During a day, the Sun rises in the east moves across the sky and sets in the west. But because the Sun appears as a disk, not all of the Sun is seen to rise or set at once. Therefore, a definition of rising or setting must be precise. To simplify things, the Earth is considered to be a perfectly smooth ball and the visible boundary between the Earth and the sky as being the horizon. Sunrise and sunset are then defined as the times when the upper edge of the disk of the Sun is on the horizon. In other words, sunrise occurs when the top of the Sun first peeks above the horizon and sunset occurs when the last bit of the Sun disappears. The times cannot be precisely determined since, in practice, the actual times depend on such things as weather and local topography.

Now, before sunrise and after sunset, there is a period called twilight. During these times, light from the Sun located just below the horizon is scattered by the atmosphere and is still visible. Outdoor activities are still possible during these times and it is useful to have a way of defining when lights should be turned on.

Civil twilight is defined as the period in the morning and the evening when the centre of the Sun is six degrees below the horizon. These are the limits at which light is still sufficient to distinguish objects. At this point, you would have to turn on the lights if you are having a bar-b-que outside. Nautical twilight is when the centre of the Sun is 12 degrees below the horizon. At this point, you can no longer distinguish the horizon. And astronomical twilight is when the centre of the Sun is 18 degrees below the horizon and there is no scattered sunlight at all visible.

Which brings me back to my original question of why twilight is so different between the tropics and further south.

In the tropics, the Sun is higher in the sky during the day and consequently hits the horizon at a steep angle. The centre of the sun, therefore, gets to 6 degrees below the horizon, and hence the end of civil twilight, reasonably quickly. The further you are from the equator, the lower the Sun is in the sky and it strikes the horizon at a shallower angle. Since it takes longer to get to six degrees below the horizon, civil twilight lasts longer.

On one of my trips around Australia, this was dramatically demonstrated as I sat on the beach at the tip of Cape York. Being the most northerly point of mainland Australia meant I was well and truly in the tropics. As I sat watching the sunset, I was amazed at how quickly it got dark. I was used to civil twilight lasting for around 30 minutes, but at Cape York, it only lasted about 10 ten minutes before the stars became easily visible. I had my first ever glimpse of the small, faint Andromeda Galaxy just minutes after the Sun went below the horizon.

One year

Continuing with the time-based theme, what do we mean by a year? Essentially, one year is the time it takes the Earth to make one complete orbit of the Sun. But, as with all other astronomically based time measurements, things are not so straight forward.

There are several different ways of defining a year. If we could get above the solar system and look down on it from a sufficiently far off distance, a Sidereal Year is the time it takes the Earth to revolve around the Sun back to an exact starting point in space. The precise length of this year can be determined using the stars and is 365 days 6 hours 9 minutes and 9 seconds long.

Human activity, however, tends to be ruled more by the seasons than the stars. The time it takes the seasons to cycle through and start again is known as the Tropical Year and is the year commonly used in our calendars. The length of a tropical year is 365 days 5 hours 48 minutes and 45 seconds. The 20 minute difference between the two years is due to a slow wobble in the Earth's axis.

Since the commonly used tropical year isn't an exact multiple of a day, a bit of fiddling was required to come up with a usable calendar. With the actual year almost equal to 365¼ days, it was decided by Julius Ceasar in 45 BCE to make a Civil Year an exact number of days, typically 365. Of course, this means the extra quarter day slowly accumulates and the calendar gets more and more behind. Every fourth year an extra day is put into the calendar to bring it back in line. This is known as a Leap Year.

The rule for calculating whether a year is a leap year is relatively simple. If the year is divisible by four, then it is a leap year, unless it is a century, when it must be divisible by four hundred. As a result, the year 2000 CE was a leap year, whereas 1900 and 2100 are not leap years, even though they can be divided by four.

Even this is not exact. The current calendar accumulates an error of one day in about 3,300 years, but we'll worry about the extra day when the time comes in the year 4909 CE.

Seasonal changes

Between the hot days of Summer and the cold nights of Winter, the Earth's weather continually changes throughout the year. These seasonal changes are not generated by the Earth, but by the passage of the Earth around the Sun.

On its yearly orbit, the Earth moves with its rotational axis tilted by about 23½ degrees to the plane of its orbit. Since the rotational axis always points to the same spot among the stars, sometimes (as the Earth revolves about the Sun), the south pole is tipped towards the Sun, and sometimes (six months later) the north pole is tipped towards the Sun. In between neither is tipped towards the Sun. It is this changing tilt that causes the seasons.

Now, when the southern hemisphere tilts towards the Sun, we have Summer. Not because we are necessarily closer to the Sun but because the Sun appears higher in the sky and the sunlight falls more squarely onto the surface. That means a square metre of sunlight in space corresponds essentially to a square metre of sunlight on the ground. This concentrated energy causes the earth and the water it falls on to heat up and hence gives us summer weather.

Six months later when the southern hemisphere is tilted away from the Sun, we have winter because now the Sun is lower in the sky and the sunlight is falling obliquely onto the surface, spreading the energy over a larger area. In other words, our square metre of sunlight in space now falls over much more than a square metre on the ground. With the energy from the sunlight now spread over a larger area, the land and water cool down and we have colder weather. In between these two extremes we get Autumn and Spring.

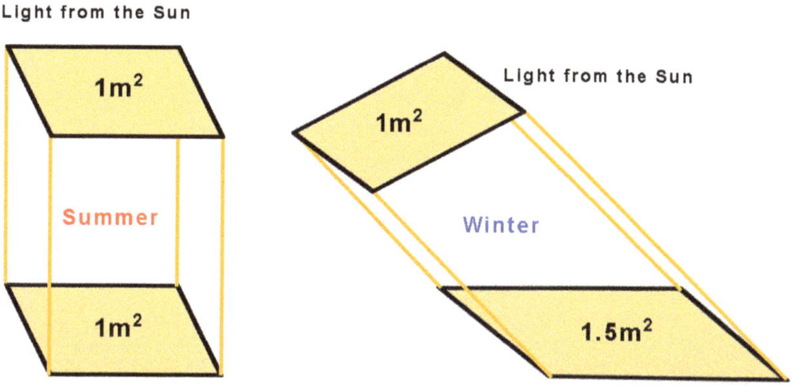

Sunlight striking the Earth at different angles throughout the year determines the seasons.

The changing tilt of the Earth relative to the Sun accounts for why the maximum height of the Sun at noon changes throughout the year. It also explains the origin of the Tropic of Capricorn and the Tropic of Cancer. The two tropics are the furthest places south and north that the Sun will ever appear directly overhead during the year. At the middle of summer, the Sun will be overhead at noon only on the tropic of Capricorn, and in the middle of winter, the Sun will be overhead at noon only on the tropic of Cancer. In between the two tropics, the Sun will be overhead two times throughout the year. Outside the tropics, at no time. And in case you were wondering, the two tropics get their names from the fact that the Sun is meant to appear in the constellation Capricorn when it is at its most southerly point and in Cancer when it is at its most northerly point. As it turns out, the Sun is no longer in these constellations at these times, due to the slow wobble of the Earth's axis.

The maximum tilt towards the Sun by the Earth's southern axis occurs not long after the beginning of summer. So why then doesn't the hottest part of the season occur at this time, around the 21st of December? If the Earth didn't have an atmosphere or oceans it would be closer to this date. However, both of these have the effect of regulating the overall temperature we experience. In particular, it takes time for the oceans to heat up and so the peak of summer doesn't occur until quite a bit after the maximum tilt of the axis. For summer this means the hottest time is generally around late January or early February, not December. Similarly, in winter, the atmosphere and oceans take time to cool down, so the coldest part is not in June but around the end of July and August.

The four seasons are approximate periods during the year characterised by particular weather conditions. With Summer, things are generally warm. With Winter, they are usually cold. With Spring and Autumn, the weather conditions are typically in between, slowly getting warmer/cooler after the cold/warm Winter/Summer months but not quite there yet. However, if we use this definition, we could argue that Autumn, for example, could start anytime from mid-February through till April. Why don't we start it then?

Since the concept of Autumn weather has a very fuzzy definition, it is impossible to define precisely when we can say without a doubt that summer ended and Autumn began. Consequently, an arbitrarily determined start date for each season is picked for the calendar. The dates agreed upon should, of course, encompass the generally accepted weather associated with each season, but essentially a date needs to be picked. In many countries, the dates chosen for the start of each season are the equinoxes (the times of equal day and night) and the solstices (times of shortest and longest days). This gives an apparent astronomical basis for the start of each season. The problem is that these dates vary each year. They can occur on anything from the 20th to the 23rd of the month. The countries that adopt this definition, however, start the seasons on the 21st of the month, regardless, rather than altering it to fit the actual date each year. Since this is an arbitrarily set date and not strictly following the astronomical basis for determining them, why not set a date that is easier to remember and more convenient? In Australia, that is precisely what we do. All of our seasons start at the beginning of the month. A much more convenient way of doing it, especially since summer weather unquestionably begins well before the 21st of December.

Do any of the other planets have seasons? The short answer is yes, some do. Mercury, Venus and Jupiter have essentially no tilt, so it would always be Spring if it weren't for other, overriding factors such as thick clouds and a nearby star. Saturn and Neptune have similar tilts to the Earth but their seasons would last seven and 41 years respectively. Uranus is a strange case. Being almost tilted on its side means that sometimes the poles point directly at the Sun and other times the equator does. Thinking about how that affects its seasons is enough to make your head hurt.

On Mars, things are a lot more familiar. The tilt of Mars is almost identical to Earth's and the length of the day is only slightly longer. The main difference, apart from being colder, is that its year is twice as long, so the seasons are also twice as long as they are here on Earth. Otherwise, they would be strangely familiar to us. Maybe sometime in the not too distant future humans will get to experience the Martian seasons when we colonise our next-door neighbour.

Analemma

As the seasons change, so do the times of sunrise and sunset. Around the summer or winter solstice, we find the time of sunrise and sunset hardly changes from day to day while around the spring and autumn equinoxes they change dramatically.

Around a solstice, the differences in day lengths are measured in seconds, and since the rise and set times are usually given in minutes, you often see very little change in the stated times for a few weeks. But if you take note of the times the Sun rises and sets you will also notice they're not equally situated either side of noon.

If you were able to observe the Sun safely, you would notice that very rarely is it due north when your clock says it is noon. You may not think this is a big deal, but theoretically, noon is when the Sun should be halfway across the daytime sky, placing it due north at this time. In reality, the time at which it reaches this point varies throughout the year, and the reason for that lies with the Earth's orbit.

The Equation of Time describes this varying difference between noon and when the Sun is due north. Anyone who owns a sundial will be familiar with this. Sometimes it is displayed as a table of numbers used to adjust sundial time to clock time, and sometimes as a curious figure '8' shape, known as an analemma. The east/west part of the figure '8' is due to the Earth's orbit, as mention above. Since the orbit is an elliptical shape rather than circular, as the Earth moves around, it regularly gets closer and further from the sun, causing its speed to change continually. It is this changing speed that generates the east/west variation. The changing height of the Sun throughout the seasons due to the tilt of the Earth's rotation axis causes the north/south aspect of the analemma.

Exploring for yourself: Some year-long solar measurements

This observation requires more of a commitment in time than ones I will mention later on, but it does involve perhaps the simplest of instruments, a vertical stick on a flat piece of ground.

Over an entire year, you will measure the altitude and azimuth of the Sun at the same time of day once a week, preferably as close to noon as possible. Be careful to take into account any daylight saving times as all measurements should be at your local standard time. These observations eventually produce a series of shadow measurements, and from them you can calculate when the Sun is at its maximum and minimum altitude, when the solstices occur, and your latitude.

To start, find an existing vertical 'stick' such as a pole, fence post or clothesline. If none of these are available, an actual stick stuck in the ground and left there for the year will do. It isn't advisable to use a tree as your 'stick' as they grow over a year and will throw out your measurements.

You also need some way of recording your measurement. This could be anything from a series of pegs in the ground, chalk marks on a concrete slab, or a piece of paper that can be removed and put back in precisely the same position and orientation each time.

However you decide to record your measurements, each week put a mark at the end of the shadow caused by your 'stick' and record the date. There will be weeks when you cannot see the Sun due to the weather, but that's not a problem. If it is clear in the next few days, take a measurement then. It does not matter. If you wanted to be really keen, you could do it every day and end up with an exceptional result. After doing this for a year, you will have a series of marks that trace out a distorted figure '8' pattern, known as an analemma. This analemma can now be used to work out some information.

Plotting an analemma.

The top of the figure '8', when the shadow was at its longest, is the date the Sun was at its lowest in the sky, or the Winter Solstice. The bottom, when the shadow was at its shortest, is the date the Sun was at its highest in the sky or the Summer Solstice.

Measuring the height of your 'stick' and how far it was from the base of the 'stick' to the bottom of the analemma (the summer solstice) means you can now work out the angular height of the Sun on this day using some high school trigonometry. Subtracting this angle from 90 degrees and adding 23½ degrees will then tell you how far from the equator you are, in other words, your latitude.

For example, if the stick used was 100 centimetres tall and the bottom of the analemma was 19 centimetres from the base of the stick, then the angular height of the Sun would be 79 degrees. Subtracting this from 90 degrees gives 11 degrees. Adding $23\frac{1}{2}$ degrees (the difference between the Equator and the Tropic of Capricorn) means the latitude is 11 plus $23\frac{1}{2}$ or $34\frac{1}{2}$ degrees south.

Tidal influence

Twice a day, the ocean goes through a cycle of high and low tides. But what causes these variations? Generally, the difference between high and low tide is only a couple of metres at most, although in some places, because of the geometry of the land, tides can be a lot greater. For instance, Derby in the Kimberley region of Western Australia is a great place to observe the biggest tides found anywhere in the tropics.

The tidal bulges are caused by the battle between the gravitational attraction of the Moon and the centrifugal force away from the Moon due to the Earth and Moon orbiting their common centre of mass. Centrifugal force is the pseudo-force away from the centre of rotation experienced when moving in a circular motion. It is the force that keeps you firmly in your seat as you go upside down on a rollercoaster.

The differences in water heights occur when the Moon's gravity wins out and attracts the water closest to the Moon, while on the other side of the Earth the centrifugal force wins out and forms a bulge of water on the side away from the Moon. In between the two bulges, the water depth lowers as the water had to come from somewhere. As the Earth rotates, wherever you are on the Earth passes through these two high and two low tides every day.

However, if that were the entire story, you would expect a high tide whenever the Moon was directly overhead. But since the Earth rotates faster than the Moon moves around the Earth, the friction between the high tide bulge and the ocean bottom produces a lot of complicated effects. Ocean currents, sloshing effects between the ocean and land masses, and the main tidal bulge being ahead of the Moon are just some of the consequences caused by this interaction.

Friction between the tide and ocean bottom is also slowly slowing down the Earth's rotation. Currently, the length of the day increases by about 1.4 milliseconds per day per century. Since the formation of the Earth 4.5 billion years ago, the day has increased in length from a rapid six hours to the current, more leisurely, 24 hours.

Even if we didn't have the Moon, we would still have tides caused by the Sun, although they wouldn't be as large as those we have now. It is the combined influences of the Sun and Moon that ultimately generate the tides we get to see. At the time of a new or full Moon, when the influence of the Sun and Moon's gravity combine, we have large high tides and shallow low tides. At first or third quarter, when the Sun and Moon are at 90 degrees to each other, their effects cancel a bit, and the difference between high and low tide is much less pronounced.

Where are we heading?

I sometimes get asked to point out the location in the sky that the Earth is heading. Everyone knows we are moving in space, so what direction is it? Usually, the follow-up question is to ask how fast we are getting there.

Most people expect me to point to a particular spot in the night sky. But the Earth's motion through space depends on all sorts of things, and it is not as easy a question to answer as you might think.

If we talk about the Earth's motion with respect to the Sun, we always seem to be travelling toward a point about 90 degrees west of the Sun. We do this at an average speed of about 30 kilometres per second. But that's just being pedantic, as I know it is not the question people are trying to ask. If we talk about what direction the Sun, and by association the Earth, is moving with respect to the centre of the Milky Way galaxy, then, due to our orbital motion around the galactic centre, we are heading towards the constellation Cygnus, the swan, at about 250 kilometres per second. This is usually what people want to know when they ask the question.

But the Milky Way is a tiny part of a vast universe. So in the grand scheme of the entire cosmos, where are we headed? That's a tough question to answer, since, in order to determine a direction, we need a frame of reference against which to measure our motion. With the Earth around the Sun and the Sun around the Milky Way, it is relatively easy to find a fixed background against which to measure. But if we look at the

Milky Way's motion, we need to have an all-encompassing background that permeates the entire universe if we are to work out how we move with respect to it.

Fortunately, cosmologists have discovered something that fulfils exactly those requirements, the remnant radiation from the Big Bang itself. The cosmic microwave background radiation, which dates from soon after the Big Bang and covers the entire sky, allows us to work out in what direction we're ultimately moving. Against this all-pervading cosmic glow, the Sun is travelling at about 370 kilometres per second toward a point on the border between the constellations of Crater (the cup) and Virgo (the virgin).

CHAPTER 14

Lunacy

Lunatics

Does the Moon affect our daily lives? It seems there will always be some people wanting to believe there is a cosmic connection between the Moon and us and that it continually affects people. A classic example is the interest in so-called Super Moons.

This pairing of a Full Moon at the time of its closest approach to the Earth occurs on average about five times a year, so they aren't particularly rare. Some people are concerned with the effects this might have on the Earth, even attributing coincidental natural disasters to Super Moons. The problem with such connections is that generally, they match the Super Moon with an event that occurred near the same time. But unless it happened at precisely the time the Moon was at its closest, there's no point trying to correlate the two.

Severe storms, earthquakes and volcanic eruptions are also supposedly associated with the Full or New Moon, and, as a general rule, will occur within three days either side. But if you think about it, that's not bad odds for an association. Out of a maximum 31 days in the month, the prediction is that sometime in 12 of those days somewhere in the world, some natural event will happen. With these odds I'm surprised there aren't more associations made than there are.

Looking at possible effects on humans, in medieval times the Moon was blamed for all sorts of things. For instance, making people mad (the origin of the words lunatic and lunacy) or turning them into werewolves. Though there may be some detrimental effect on your health due to a disturbed night's sleep caused by the brighter light of a Full Moon, I'm not so sure it influences people to quite that extent.

Therefore, the argument is generally more focused on the gravitational or tidal effect on the human body. Although the Moon's gravity does create tides in the oceans, even lakes are too small to be affected, so there is no chance of tidal effects occurring in a human body. Plus, the Moon passes over us every day, so why doesn't the Moon's gravity affect us each day? In fact, a person standing next to you exerts more of a tidal force than the Moon. And to get really picky, any effect is entirely overwhelmed by the natural compression of the human body while standing up. The lunar tides stretch you less than you shrink from Earth's gravity.

Lunar features

As our only natural satellite, the Moon is possibly big enough compared to the Earth to consider the two of them a double planet. It is also the only other object apart from the Earth humans have so far walked on, the last time in December 1972.

Even if you are not one of the lucky astronauts who saw it up close, a casual glance with your eyes will show that the surface of the Moon is not featureless. It is covered by numerous spots, lines, mountains and large dark areas.

The main features we can see on the moon come in three distinct groups. Irregular blotches on the surface (craters, mountains, faults) are the most common. The large dark regions are known as 'seas' and 'oceans' and are the easiest to see with just your eyes. And the most-subtle group are the bright rays that emanate from the larger craters.

The dark regions are large, relatively flat areas on the surface formed when the surrounding rock was turned molten in some cataclysmic event. Any low lying areas were filled in by this molten rock to produce the massive plains that have subsequently been relatively undisturbed. Over the aeons, they have turned darker through the slow weathering caused by the solar wind and small meteorites. Since the Moon doesn't have any liquid water, they're not actually seas, but when people first looked at the Moon, that's what they thought they looked like and the name has stuck. Without a doubt, the most famous 'sea' is the Sea of Tranquillity, as it was where the first human set foot in 1969.

A lot of people see a face made out of the seas and oceans when they look at the full Moon. I have to confess I struggle to see a face, and instead, find it easier to see a rabbit made out of the dark patches. If you look at the Moon and see the rabbit on its left-hand side, it was at the base of the right-hand ear that Neil Armstrong left his famous footprints.

But the most common, most interesting and perhaps most desired features to look at are the craters, mountains and other faults on the surface. Occasionally the Moon is struck by a big meteor, and the collision produces a crater. In the process, a large volume of material is often ejected and splashes over the lunar surface. This material is brighter than the surrounding areas as it has come from under the surface where it has not been subjected to the same weathering. It then appears as a display of bright rays emanating from the crater. The two most prominent craters on the Moon are called Copernicus (93 kilometres across) and Tycho (85 kilometres across). Both of these are notable for the rays that emanate from them. The Earth also has impact craters, but they're not quite as numerous, as erosion tends to get rid of them fairly quickly.

Since the darker seas and the brighter rays have no fine detail, only contrast between light and dark, the best time to see them is at a Full Moon, when the sunlight is shining straight down on them and there are no shadows to interfere with the view. But if you want to see details, then a First or Last Quarter moon is best. At this time, the Sun is off to the side, creating shadows that outline the craters and other features, making them a lot easier to see.

Exploring for yourself: Features on the Moon

Even with just your eyes it is remarkable how many features can be seen on the Moon. Binoculars or a telescope make it even easier. Starting a few days after New Moon and continuing until Full Moon, use the map on the next page to find as many of the features as possible as the terminator (the line between day and night) passes over them, including the manned landing sites on the Moon. Around First Quarter Moon is an ideal time to see the 'Rabbit in the Moon'. There are actually two different rabbits, so try and find both.

The Darkside of the Moon

Gazing at the full Moon won't make you a lunatic, but if you stare at it often enough, you may realise that you only ever see the same features on its surface. Why don't you ever see them change? It all has to do with the orbit of the Moon and a trait common among planetary satellites. The Moon is in something known as synchronous rotation around the Earth. That is, the Moon rotates once on its axis in the same time it takes to go once around the Earth. This is a consequence of the tidal effects of the Earth's gravity slowing the Moon's rotation until it has become locked with the Earth. It wasn't always like this.

Early on in its history, the Moon was a lot closer and used to rotate on its axis much faster, but the gravity of the Earth slowly but surely slowed the Moon's rotation down. It is now locked into this synchronous rotation and will not slow anymore.

Because of its synchronised rotation, when we look at the Moon from the Earth, the Moon always keeps the same face pointed towards us. This implies that you would expect to see only 50 per cent of the Moon's surface and consequently only the same features all the time. But in reality, due to an effect called libration, we get to see about 59 per cent of the surface.

Since the orbit of the Moon is an ellipse, as it goes around the Earth its speed varies, moving more rapidly when nearest the Earth and more slowly when furthest away. Since it still spins on its axis at the same rate, but its speed around its orbit changes slightly, the Moon appears to wobble a bit in an east-west direction, and this allows us to see a little bit more of the Moon in these directions.

The tilt of the Moon's orbit causes the other libration effect. Since its orbit is tilted by about five degrees compared to the Earth's orbit around

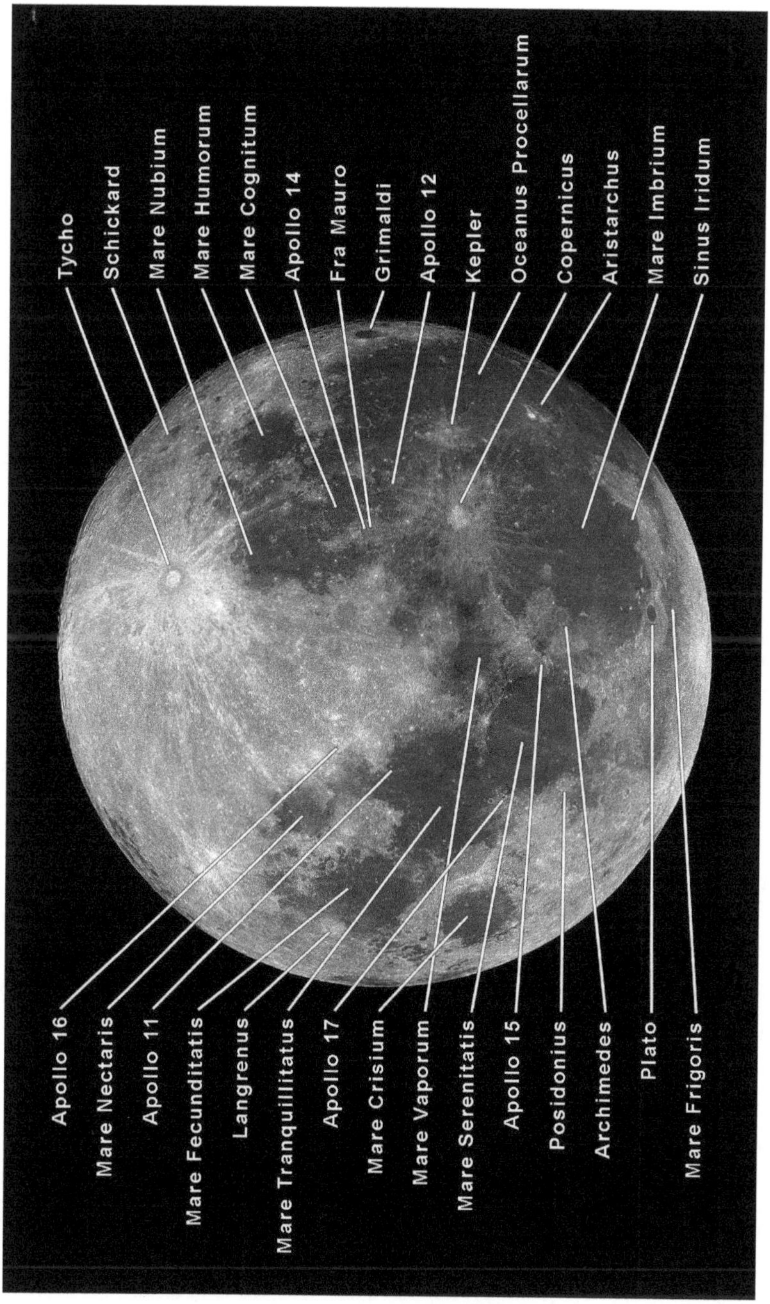

Some of the easier to see features on the Moon.

the Sun, we can see a little over the south pole of the Moon during one half of its orbit, and during the other half, we can see a little over the northern limb.

Combined, these libration effects mean we see more than just half of the Moon's surface. Keep an eye on some of the easy to observe features near the edge of the Moon for a week and you will spot some of this wobbling.

Seeing the same side all the time has led to the idea that there is a dark side of the Moon. Any object in space will have one side lit up by the Sun and one side in darkness, so this is true, and as I type this, I am currently on the dark side of the Earth. That is, at the moment it is nighttime where I am. But even though we only ever get to see one side of the Moon from here on the Earth, that side is not always lit up.

The Moon does rotate, just very slowly. Over a month, this means that every side of the Moon will, at some stage, be facing the Sun and lit up and sometimes not facing the Sun and in darkness. When we see a Full Moon, it means the side facing us is fully lit up, and the far side of the Moon is in darkness. With a New Moon, the side we see is entirely in darkness, and the far side is lit up. So, despite what Pink Floyd tell us, there is no side of the Moon perpetually in darkness. A better phrase for the unseen part of our lunar neighbour would be the Far Side of the Moon.

If the dark side of the Moon is where there is no sunlight, then why is it that when the Moon appears as a very thin crescent from here on the Earth it is possible to see some details on the unlit portion of the Moon? One of the greatest geniuses of all time, Leonardo da Vinci, was the first to explain this phenomenon. He correctly described this spectacle as the glow from a very bright Earth shining onto the dark side of the Moon and then faintly back to us. We consequently call this phenomenon 'Earthshine'.

When the Moon is just off from being directly between the Earth and Sun the dark side still sees a very bright Earth in its sky. This means the dark side isn't genuinely dark at these times. It is similar to how a full Moon in our night sky makes the night quite well lit. As the Moon continues to move around in its orbit, it sees more and more of the Earth's unlit side, so there is less light from the Earth illuminating its dark side, and we no longer get to see any Earthshine.

Before I leave the subject, years ago I had the privilege of seeing

Leonardo's original manuscript known as the Codex Leicester. It is the document in which he first describes the phenomenon of earthshine, among a multitude of other ideas he had. It was on display at Sydney Observatory for a week, complete with an armed guard, and every day I would take time to stand and marvel at the genius of Leonardo da Vinci over 500 years ago.

Phase brightness

The Moon is a lot brighter when it is half full (First or Last Quarter) than not lit at all (New Moon), but is it then correct to say that it will be twice as bright when fully lit at Full Moon? Like most things in astronomy, it isn't as clear cut as it might seem at first glance. If you carefully measure the Moon's brightness at each of these phases, you find that it can be up to 10 times brighter when it's full than when it is at First Quarter. There are two reasons for this.

Firstly, from our viewpoint on the Earth, when the Moon is full, sunlight is shining straight down onto the Moon's surface. That means there are no visible shadows. However, when the Moon is at First (or Last) Quarter, the Sun is off to the side, and there are lots of shadows. These have the effect of darkening the Moon's surface. In fact, at Full Moon, there is more than twice as much lit up surface visible than at First Quarter. Secondly, over millions of years, meteorite impacts and the harsh radiation from the Sun have slowly eroded the top few centimetres of the lunar surface into an extremely fine powder, similar to ground flour. Unlike most objects, which scatter light in all directions, this powdery lunar soil has the peculiar property of tending to reflect light directly back to the source. So, when the Moon is half full, and the Sun is off to the side as seen by us, the lunar soil reflects most of the light back toward the Sun, which is unfortunately away from us. But when the Moon is full and the Sun is directly behind us, sunlight that hits the Moon once again reflects back towards the Sun but this time we're in the same direction, and we get to see more of it. Combined with the lack of shadows, this effect makes the full Moon much brighter than you might expect. But no one notices this dramatic change in brightness because the human eye and brain are not good at comparing things separated by an extended period of time.

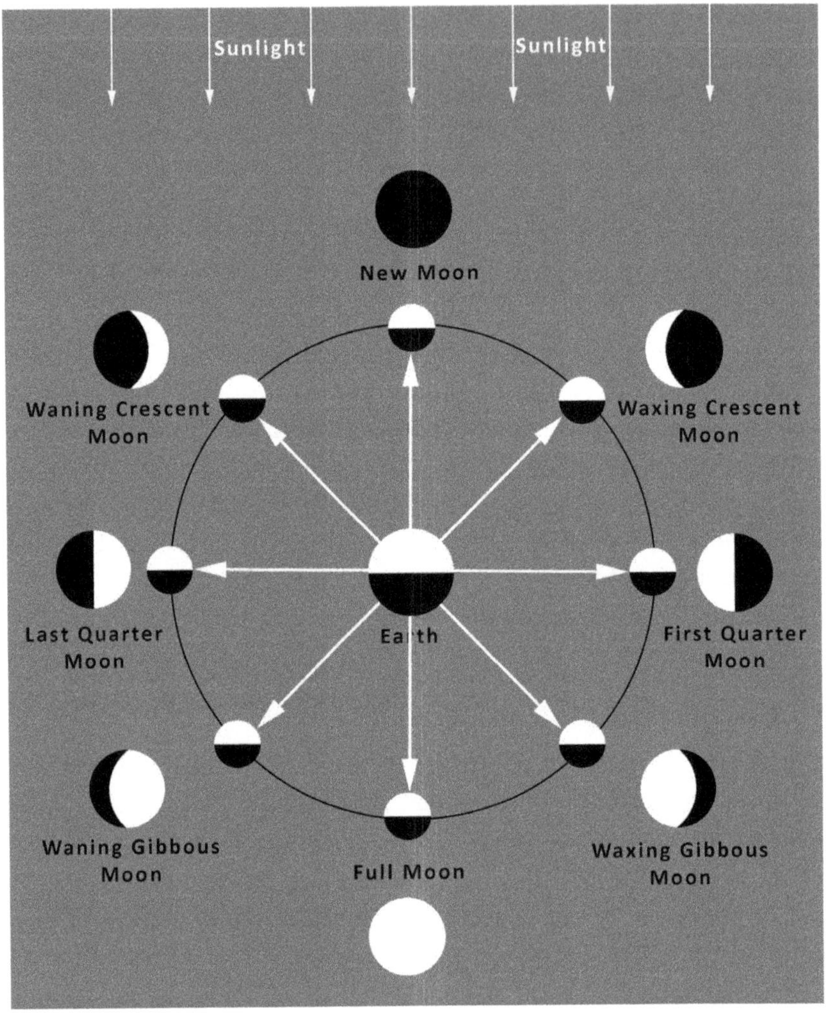

Moon phases and how they look from the Earth.

Exploring for yourself: Phases of the Moon

Observe the Moon every night for two weeks at the same time (start about half an hour after sunset on the first night) and from the same place and make a note of the shape and direction of the Moon. The Moon will go from being a thin crescent shape to a Full Moon and each night appears further to the east.

Exploring for yourself: Libration of the Moon

The Moon only shows one side to the Earth, but due to libration effects we get to see a total of about 59 per cent of the Moon's surface. This wobbling due to libration can be seen easily if a prominent feature near the edge of the moon is observed for a few weeks. One of the easiest naked eye features to use is the dark area known as Mare Crisium (the Sea of Crises – good thing no Apollo craft landed there!). Starting just after a New Moon, observe the Moon every night until Full moon, paying attention to Mare Crisium, the round sea on the edge of the Moon. After two weeks, the view of the Sea of Crises will have changed significantly.

Exploring for yourself: Earthshine

Starting a day or two after a New Moon, observe the Moon every night for about a week. In the beginning, the unlit part of the Moon is still faintly visible, but after a few days, this faint glow of earthshine will slowly disappear until the dark side of the Moon does now appear dark.

Apogee and perigee

As the Moon goes around the Earth, it does so not in a perfect circle but a slightly flattened, oval shape known as an ellipse. This means the distance from the Moon to the Earth varies slightly over the time it takes the Moon to do one complete orbit of the Earth. Consequently, sometimes the Moon is a bit closer to the Earth than it is at other times, and sometimes it is a bit further away from the Earth.

The closest approach of the Moon to the Earth is called perigee, and

Earthshine.

the furthest point in the Moon's orbit is known as apogee. Between the two distances, there is a difference of about 50,000 kilometres, although the exact amount changes from month to month. This difference in distance creates one of the most striking effects that escape notice by virtually everyone, simply because the eye and brain can't compare events over an extended period of time.

If we look at the Moon when it is full and at perigee and then look at it when it is full at apogee, what differences can be seen? The first is that the Moon appears about 23 per cent larger in area at perigee than at apogee. Due to this increase in size and its closer proximity, the Moon also appears approximately 30 per cent brighter than at apogee, or 12 per cent brighter than an average full Moon. This increase in brightness is difficult to notice due to the way the human eye processes variations in brightness. Additionally, these differences are also difficult to notice because they occur so far apart in time. If, however, you were to photograph the Moon at each of these times, the variation in size and brightness becomes readily apparent.

Blue Moon

For some strange reason having a second full Moon in the same calendar month is known as a Blue Moon. Surprisingly, this calendar meaning does not, as one might imagine, go back into ancient folklore. The earliest reference to a Blue Moon appears in the sixteenth century, but back then it was more a way of saying something was an obvious absurdity, rather than meaning a rare event.

When the volcano Mount Pinatubo in the Philippines erupted in June 1991 and spewed enormous amounts of volcanic dust and ash into the atmosphere, we saw blue suns and moons. Dust from sandstorms and smoke from bush fires can also turn the Moon blue(ish). It is from unusual conditions such as these that the expression 'Once in a Blue Moon' probably evolved, implying a rare event. However, somehow it developed the more specific meaning of a second full moon in a calendar month, and nobody knows exactly why or when.

The first known reference to a calendar Blue Moon is in the Maine Farmer's Almanac of 1937. The Blue Moon mentioned here was a seasonal one. Each season had three full Moons, each clearly defined by name (Harvest Moon, Hunter's Moon, etc.). According to the almanac, when a fourth Full Moon occurred in a season, the third one was called a Blue Moon. I have no idea why it was the third and not the fourth Full Moon. This definition was then picked up by other publications that misinterpreted this seasonal meaning and changed it to the current definition of a second Full Moon in a calendar month.

Is the calendar Blue Moon rare? Not really. Since the lunar month (from Full Moon to Full Moon) is 29 days in length and the average calendar month is 30 days, a month can often have two Full Moons. February is the only month that can't accommodate a Blue Moon, as it is shorter than 29 days.

Rings around the Moon and Sun

Generally, stargazing requires a beautiful clear sky. Clouds are the bane of an astronomer trying to see the stars. However, if the clouds happen to be a thin even cover, the Moon desperately shines through and in the process becomes surrounded by a ring of light. This ring is not directly associated with the Moon itself but merely a display within the Earth's atmosphere, similar to that of a rainbow.

When water droplets in the upper atmosphere experience cold enough temperatures, ice forms. It is this thin layer of ice crystals that produces not only the hazy clouds but also the observed ring. When the light coming from the Moon passes through this layer, the ice crystals bend the light into a circle surrounding the Moon.

The bright ring is always at a distance of 22 degrees from the Moon due to the internal reflections of light within the ice crystals, but this is not the only ring that can appear. Diligent observers may notice a second ring at a distance of 46 degrees. Unfortunately, this second ring is quite faint and to be seen requires almost perfectly stable atmospheric conditions over a large area of the sky. Never-the-less it is worth the effort to look for the next time a ring appears.

Inside the main ring, the sky appears slightly darker than outside the ring. This might seem odd since the inside is closer to the Moon, but it turns out no light is reflected towards the inside of the ring, and hence it is not as bright as it should be. This is highlighted by the ring having a sharp inside, but a fuzzy outside, edge. The effect is also exaggerated through contrast. The space between the bright Moon and bright ring appears slightly darker than it really is.

Incidentally, rings are not only the domain of the Moon. If the right conditions occur, a ring can also form around the Sun. These are not noticed quite so often because people sensibly don't look in the vicinity of the Sun. It also helps to be wearing sunglasses, as they cut down on some of the glare from the bright sky.

Would we survive if we had no Moon?

What would it be like if the Moon didn't exist? A few questions immediately spring to mind. Where would romance be without a Moon to sing about in love songs? How would lunatics and werewolves survive with perpetually dark nights and nothing to influence their behaviour? Would humans have ventured past Earth orbit if we had to travel further than the Moon to do so? And it is interesting to think about how the world's calendars would have evolved without the month.

From an astronomical viewpoint, looking at lunar eclipses and the circular nature of the Earth's shadow on the Moon first led astronomers over 2,000 years ago to believe the Earth was spherical. And without the Moon, solar eclipses don't occur. These gave us our first views of the Sun's corona and revealed that there was more to the Sun than first meets the eye.

Of course, the main differences would be physical. Without the Moon, there would be more collisions between the Earth and rocks in space, and even though we would still have tides, they would be caused by the Sun alone. These would only be about one third as high, and there would be no king tides.

The consequences of only having solar tides would be dramatic. The tidal pull of the Sun alone would not have been enough to slow the Earth's rotation to anywhere near the amount the Moon has. When the Earth was formed, a day lasted just six hours. After all this time, the length of the day would still only be around eight to ten hours long, rather than the 24 hour day the Moon has created. Imagine what life would be like if our day had only four to five hours of sunlight?

Also, without the Moon, the tilt of the Earth's axis would be unstable, and it is the tilt that is responsible for the seasons. Combining the effects of a short day and the axis moving, the Earth would experience enormous shifts in weather with raging daily winds, cyclonic storms of unbelievable severity and massive extremes in climate change. Under this environment, it is uncertain how evolution would play out, or even if it could function at all. Life itself may never have happened.

We needed a meteor impact 65 million years ago to wipe out the dinosaurs and clear the way for mammals. But we may also need to thank the cataclysmic impact that created the Moon for there being any life on Earth at all.

CHAPTER 15

Solar System Shenanigans

Finding planets

As I was growing up, I use to look into a clear night sky and wonder if any of the twinkling points of light might be planets. Like most people, I wanted to find the big ones, Saturn and Jupiter, but the one I most wanted to see was elusive Mercury. Don't ask me why, I just did. I spent hours trying to identify them. My method of planet hunting was sound, or at least it was in the mind of a young teenager, as I had read somewhere that stars twinkle and planets don't, so I figured if I looked around the sky for a point of light that wasn't twinkling, then that would be a planet. Sadly, I had been led astray. Everything twinkles, including the planets, so I never did find any of them until years later when I learnt how to locate them correctly.

So, if twinkling status won't do it, how can you tell which point of light is a planet and which is a star? Few printed star maps will have the planets marked on them because they move so fast compared to the stars. If you were to mark the planets on your map, it would only be suitable for that year. Leave them off, and the star map is good for a thousand years. So star maps are out if you want to find the planets.

The word 'planet' comes from the Greek for 'wandering star' since the planets move compared to the background stars. So one way to find a planet would be to pick a point of light in the sky, watch it for a month or two and see if it moves compared to the stars. If it does, chances are it is a planet. If it doesn't, pick another one. Alternatively, you could narrow down your search by looking up which constellation the planet is meant to be in and once again pick a bright point of light within the constellation, watch it for a month or two and see if it moves compared to the stars. If it doesn't, pick another one. Obviously, this approach takes a lot of patience.

Another way to confirm you have found a planet is to look at it with a telescope, or in a pinch, a good pair of binoculars. Through even the biggest telescope in the world, a single star will look like a pinpoint of light. But through any telescope, no matter what size, a planet will show a disc. It may not be a big disc, but it will definitely be more than a pinpoint and tell you definitively that it is a planet, not a star.

An easier way would be to use a program or an app that tells you when a particular planet is visible and where to look. You can also then use an app to check whether you have the right point of light, but that somewhat defeats the satisfaction and sense of achievement you get from being able to find it for yourself.

Whichever way you choose to locate them, Venus and Jupiter are the easiest to find. They are the brightest points of light in the night sky after the Moon. Everyone who has ever looked up at night will have seen them, whether they knew what they were looking at or not.

Saturn is not quite as bright as Venus or Jupiter. It just looks like a bright yellowish star, but if you know where to look, it is easy enough to see. The same with Mars, although it is even easier as it has a distinctive reddish hue to it.

But with Mercury, the remaining naked eye visible planet, very few people have ever seen it or at least realised that's what it was. Since Mercury never appears far from the Sun, it tends to be lost in the Sun's glare most of the time. It is, however, visible for brief periods throughout the year just after sunset or before sunrise when it appears at its greatest distance (called Greatest Elongation) from the Sun.

That leaves Uranus and Neptune to complete the set. Unfortunately, there is no getting around the fact that you need a telescope to see these two.

It took me a few years, but I have now seen all seven planets in our solar system (eight if you count the Earth) and know how to find them whenever I want. And before anyone asks why it took me so long to achieve this feat, remember, I started looking in an age before readily available computers and the internet. Finding information on where the planets were located was a lot harder back then!

What is a planet?

For thousands of years, very little was known about the planets other than they were objects that moved in the sky relative to the background stars. Today, lots of newly discovered large objects in the outer regions of our solar system challenge our definition of what it means to be a planet. You would think it is easy to define a planet - a large, round body. But difficulties arise when you start to ask just how large and just how round an object should be before it becomes a planet.

Over the years the host of objects deserving the title of planet has changed numerous times. When first discovered, the moons of Jupiter were proposed to be planets but were eventually described as moons, even though some of them are larger than the planet Mercury. The discovery of Ceres, Pallas, Vesta and Juno in 1801 had them temporarily hold the status of planets until a lot more asteroids were found and they relinquished their title. And in 1930 Pluto was discovered and initially given the title of planet.

Pluto, however, is not like the other planets and in the last few decades, numerous other objects, similar to Pluto in size, composition, distance and other properties, have been found. These discoveries raised the question of whether or not they too should be considered planets or reclassified into a new class of object in the same way the asteroids had been.

Late in 2006, the International Astronomical Union decided that rather than promote all of them and have an ever-increasing list of planets, they would finally define what it means to be a planet. A body is now considered a planet if it is in orbit around the Sun, has sufficient mass for self-gravity to form a nearly round shape and has cleared the neighbourhood around its orbit.

Even though the spectacular images sent back by the New Horizons spacecraft show Pluto looking very planet like, it regrettably fails on this last point. With its similarity to the other newly discovered objects it, therefore, had to be demoted. We now know it isn't even the biggest of these newfound objects. Had Pluto been discovered today it would probably not have obtained the status of a planet.

There are now only eight planets (Mercury through to Neptune), some dwarf planets (Pluto, Ceres, Eris, Haumea and Makemake) and numerous small bodies in our solar system.

Daytime planets and stars

The stars and planets are still up there during the daytime, so why aren't they visible? Why are we only able to see the Sun and the Moon?

Since the Sun is so bright, whenever it is visible its immense amount of light is scattered by the atmosphere and we have daylight. The Moon can be seen through this dazzling display simply because of its proximity to the Earth. Being close makes it appear large enough and bright enough to shine through the glare of the daytime sky, but everything else in space only ever appear as pinpoints. With the immense light given out by the Sun scattered by the Earth's atmosphere rendering the whole sky bright and blue, this uniform spread of brilliance overwhelms our eyes, making it difficult to see the small, faint stars and planets within it. Not only are they easily lost in the bright background, but the uniform blue of the sky robs your eyes of anything specific to focus on. This means the light from the unfocussed star is spread out, making it even harder to notice a tiny smear of light on the bright background. In addition to that, the potential number of stars you can see with your eyes during the day is limited to a handful of the brightest ones. So, the biggest problem is knowing exactly where to look and when.

The easiest way to find something during the day, therefore, is to have some help locating it. Every so often, the Moon and another bright celestial object such as a star or planet will be close together. Since the Moon is easy to find and focus on, it means the general location of the other object is also easy to find. Wearing sunglasses also helps with the brightness of the atmosphere. If you find it difficult with your eyes, try giving it a go with binoculars first, as they make it easier. When you find it, have another go with just your eyes. The funny thing is, once you find the star or planet with your eyes you will wonder how you ever missed it. However, look away then back again and it will seem to have disappeared and you will have to start again.

Very few people have seen a planet or star during the daytime, so why not be the first among your friends to do so.

The Sun

The Sun and planets formed about 4½ billion years ago from a cloud of interstellar gas. This cloud gradually condensed to form a massive ball of gas (mostly hydrogen and helium) that, squeezed by its own gravity, slowly grew hotter and hotter. Finally, the temperature at the heart of the young Sun reached 10 million degrees, setting off nuclear fusion, and the Sun was born. Without that energy being radiated out into space, there would be no life on Earth, so thankfully, there is enough hydrogen remaining in the Sun's core to keep our star shining for about another five billion years.

Our Sun is a star, just like the stars that twinkle in our night sky, but because it is much closer to Earth than any other star, it appears very large and bright. The Sun is also much too hot for anything to land on it, or even get near. If you could stand or float on the Sun's shifting, glowing gases, you would not be able to move because its intense gravity would make you weigh about 30 times what you do here on Earth. From the surface, you would see huge flares of hot gas shooting thousands of kilometres into space and the Sun's entire surface would boil like a pot of hot soup.

Like the Earth, the Sun consists of a series of layers. At the very centre of the Sun is the core. Here, the hydrogen is so tightly packed, and the temperature is so hot that individual atoms overcome their natural tendency to avoid other atoms and ram into each other, forming heavier helium atoms and releasing energy in the process. This energy then takes thousands of years to make its way to the upper layers and out into space.

Heading outwards from the Sun's core, we come to the Radiative Zone. The energy generated in the core radiates outward through this layer and into the Convection Zone. As the name suggests, the Convection Zone is where the gas now transfers the slowly cooling energy from the Radiative Zone through the process of convection, similar to what occurs with a pot of boiling water. Pictures taken of the top of the Convection Zone show large bubbles of hot gas bubbling up from deep inside the Sun.

Almost all of the visible light from the Sun comes from the next layer, the Photosphere. Although it is very hot (about 5,500 degrees Celsius), the photosphere is much cooler than the Sun's inner layers. Even cooler are dark blotches called sunspots that appear in the photosphere. Most of these giant magnetic storms are larger than our Earth and the number of sunspots increases and decreases over an 11 year cycle.

Because it is very thin and the source of the light we see, the photosphere is what we generally think of as the edge of the Sun. But it isn't the final layer. Similar to the way the Earth doesn't stop at its rocky surface, the photosphere also has a layer of atmosphere above it. The Sun's next layer is called the Chromosphere. Huge flares and loops of hot gas shoot into the chromosphere. Extending tens of thousands of kilometres above the Sun's surface, these flares shoot electrically charged particles into the solar system, disrupting communications and causing the aurorae when they reach Earth.

Finally, the outermost layer of the Sun is its Corona. Although it is extremely hot, gas in the corona is spread very thinly, so the only time we can see the corona with our eyes is during a total solar eclipse.

Exploring for yourself: Observing sunspots

Firstly, a good rule to always keep in mind is never look directly at the Sun unless you are certain it is safe. The good news is that since sunspots are dark blemishes on a bright disc there are ways to observe them without ever looking directly at the Sun.

The simplest way is to punch a pinhole in a piece of cardboard, hold it up towards the Sun and shine the spot of light it produces onto another piece of cardboard behind it. An image of the Sun and any large sunspots will be visible. Unfortunately, getting a focused image using this method is tough and, consequently, only the more prominent spots can be seen.

A better method for viewing sunspots is to use one side of a pair of binoculars to do the same thing. This is my preferred method as it is quick, flexible, and easy to set up. Making sure one side of the binoculars is covered to avoid accidentally looking directly at the Sun, then, without looking through them, point the binoculars roughly in the Sun's direction and slowly move them about until the shadow of the binoculars is as small as possible. When you cannot make the shadow any smaller, the binoculars should be pointing directly at the Sun and a bright image will appear in the middle of the binocular's shadow. Since binoculars can be focused, it is now possible to get a sharper image and see smaller sunspots. To get the best view, project the image onto a piece of white cardboard that lies within the shadow. That way, the contrast is higher, and the spots are easier to see. A telescope can be used in the same way, but unless you are familiar with using a telescope the risk of accidentally looking straight at the Sun is a lot higher and therefore more care needs to be taken.

A neutral density solar filter on the front of a 28 cm telescope.

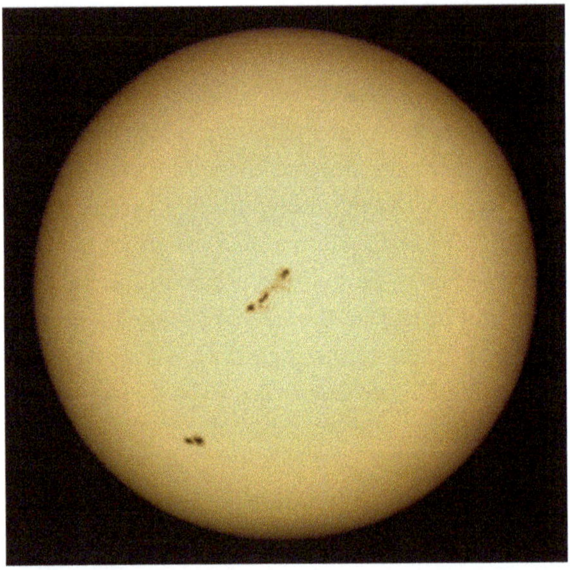

View of the sun through a neutral density filter showing sunspots.

Once an image of the Sun and its sunspots has been generated, the size of the spots can be determined by measuring the width of the Sun's image and the width of the spot. By dividing the spot's width by the Sun's width, the size of the spot as a fraction of the Sun's diameter can be calculated. Multiply this fraction by the actual diameter of the Sun (1,392,000 km), and you have the diameter of the sunspot. For reference, the Earth is only 12,742 kilometres across.

Keeping track of the sunspots over a few weeks allows the rotation rate of the Sun also to be calculated. When a sunspot appears on one side of the Sun, note how long it takes to move across the Sun's disc and disappear off the other side. Double this length of time and it will give you how long it takes for the Sun to rotate once. The time should be around 25 days, although it does vary slightly depending on what latitude the spot appears on the Sun.

The number of sunspots on the Sun's surface varies on a weekly basis, but keeping count of them over a few years will show a general trend, slowly increasing to a maximum number and then slowly decreasing until there are no spots at all visible. After a while, the number of spots start to grow again, and the cycle repeats. To see this approximate 11 year cycle does, however, mean a commitment of a few decades.

Venus

After the Moon, Venus is the next brightest object in the night sky, appearing bright due to a combination of being relatively close to the Earth, close to the Sun and covered in very thick clouds. The clouds are so thick they reflect most of the light that falls on them and being so close to the Sun means there is a lot of light to reflect. It all adds up to making Venus very bright and easy to find in the sky.

Venus is closer to the Sun than the Earth, and as it moves around its orbit it alternates between passing between the Sun and us and being on the opposite side of the Sun to the Earth. At its closest, it is only 38 million kilometres away. At its furthest, 261 million kilometres. This difference of 223 million kilometres in our view of the solar system's hottest planet means there is a dramatic change in brightness and apparent size of Venus.

On its way to passing between the Sun and us, it slowly gets closer to the Earth, and we get to see less and less of the sunlit side. In other words, Venus shows phases, just like the moon. As it gets closer, it gets brighter, but we also see less of the sunlit side. Ultimately, the brightness we see is a balance between this increasing brightness and how much of the decreasing bright side we can see. Eventually, there comes the point where the two give a maximum brilliancy before the increasing dark side of Venus takes over, and its brightness starts to decrease. Venus then passes between the Sun and us, during which time we cannot see it, and the process is repeated in reverse as it appears on the other side of the Sun and gets further and further away from us.

As Venus continues moving around in its orbit, it also reaches a point where it appears as far from the Sun as it ever gets. At most, this distance is 47 degrees and is known as Venus' Greatest Elongation. Passing this point, Venus then appears to get closer to the Sun as it travels to the other side of its orbit, eventually getting lost in the Sun's glare as it passes behind our star, before reappearing on the other side of the Sun. Depending on which side of its orbit Venus is determines whether it is visible in the evening or morning sky.

If the only motion involved was Venus moving around the Sun, then the expectation is that from the time of maximum brilliancy in the evening to the next would be the same as Venus' orbit, 225 days. However, in this time, the Earth has also moved a significant way around the Sun. Since Venus moves quicker than the Earth, it eventually catches up, but it takes 584 days to do so. That means it is 584 days between the evening maximum brilliancy and the next time it occurs.

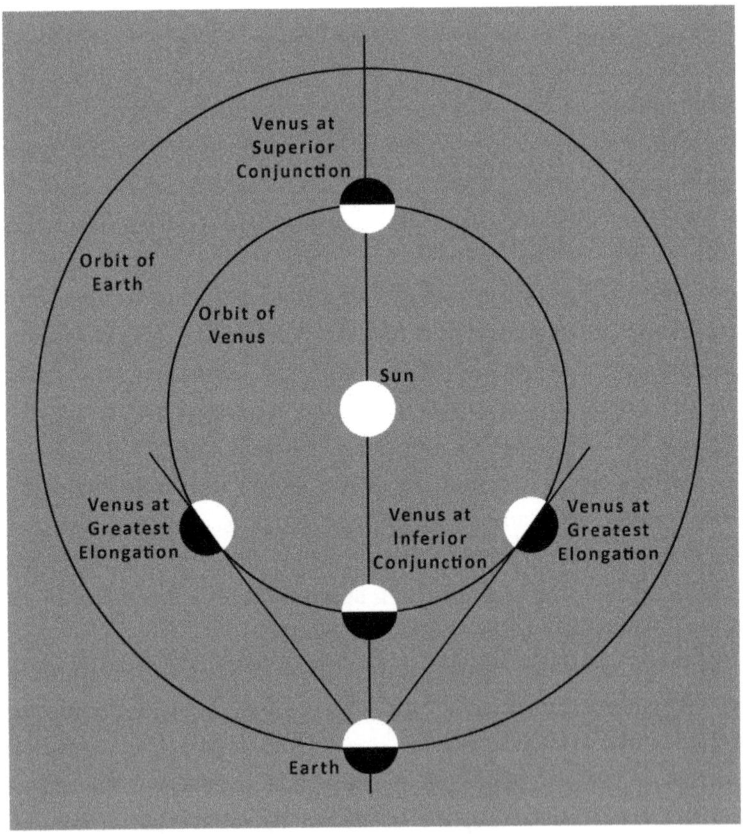

Greatest elongations of Venus.

Exploring for yourself: Venus

The changing brightness of Venus is easy to see with just your eyes. Over a few months, when Venus is visible (for convenience I suggest looking at the evening apparition of Venus), go out and have a look at the planet occasionally just after sunset. Each time you view the planet it will have noticeably changed brightness.

If you start looking when Venus is still close to the Sun then from night to night it will also slowly get higher in the sky as it gets further from the Sun. Eventually, it reaches its furthest distance (greatest elongation) before it starts to get lower in the sky and closer to the Sun again.

The phases of Venus over this period are easily visible through a telescope, but it is also possible to see them with a good pair of binoculars if you can hold them steady enough. If you can, it is well worth having a look at the phases of Venus.

One final observation to attempt with Venus is to see the planet during the day, as it is bright enough to see with just your eyes if you know where

Venus showing a crescent phase.

to look. The difficulty is in finding it. The best way to locate it is to find out when Venus will be close to the Moon and use the Moon to determine where in the sky Venus should be. A handy tip is to wear sunglasses, as they do help to dull the bright blue sky. Once found, you will wonder how you ever missed it. The first time you see Venus during the day using just your eyes is always an exciting moment, so it is worth attempting.

Mars

Over the years, Mars has been the source of many science fiction stories, and for a good reason. Ever since the invention of the telescope, we have known that Mars has the closest Earth-like conditions found anywhere in the solar system. It has polar ice caps that grow and recede with the seasons. It has large volcanoes, dry river beds and dark markings that continually change. It has a reddish colour because the rocks and sand are a little rusty, and it may still have signs of life hidden somewhere on its surface. But most of the time these features are difficult to see from here on Earth.

Mars is not a big planet, only about half the diameter of the Earth, making it appear small and hard to see. However, every 26 months, the Earth and Mars are on the same side of the Sun, and the distance between them is at a minimum. Unfortunately, due to the elliptical nature of both Earth's and Mars' orbits, not all close approaches of the two planets are the same. At their closest, they can be a mere 56 million kilometres apart, but most approaches are a good deal further apart than that. Even so, when Mars is close, it does appear bigger and brighter than usual. With a larger disc to look at, we can see some of the features listed above. Given reasonable viewing conditions, anyone looking through a telescope at these times will be able to see the dark markings on Mars' surface and maybe a bright polar cap. Keen-eyed observers might even be able to see more details, such as a channel, or even a volcano or two.

This continually changing distance has another effect on the relationship between the two planets. As the distance from Earth increases, so does the time it takes to communicate with the many spacecraft on and around Mars. Light and radio waves travel at a finite speed of about 300,000 kilometres per second, so any signal sent between Mars and Earth can take between three minutes (at closest approach) and 22 minutes (furthest distance from the Earth) to make the journey. Since rovers on the surface are driven by scientists back here on Earth, this time lag makes it difficult to avoid bumping into things, so the controllers have to be very, very careful where they tell the rovers to go.

Minimum Mars phase

Everyone at some time or another has noticed that the Moon seems to change shape. Over about a month, the Moon goes from being full to half to none to half again and eventually back to a full disc. These changes are known as phases. But why does the Moon do this and does anything else in the solar system show phases?

Apart from the Sun, every object in the solar system shines by reflecting sunlight. We see the Moon and planets only because some of the light falling onto them is reflected to the Earth and into our eyes. Because they only shine by reflecting sunlight, the side of the object facing the Sun is lit up while the side away from the Sun is in darkness. For us to see this dark side, we have to be able to get around behind the object, on the side opposite the Sun. If we can't do this, we only ever see the illuminated side. Fortunately, we do get to see the dark side of some easy to see objects. Mercury, Venus and the Moon pass between the Sun and the Earth, and this allows us at times to see their unlit side. But with the planets further out from the Sun than the Earth, we cannot get behind them, so we are never able to see any of their unlit sides.

Mars is the exception. It is close enough to the Earth that when we see it at the point in its orbit off to one side, we can catch a peek at a little bit of its dark side. We don't get to see much, but we do get to see some, which means Mars also shows some slight change in phase. At most we only ever get to see about 15 per cent of the dark side of Mars, but if you were to look at Mars through a telescope when only 85 per cent of its surface is illuminated, you would notice that it is distinctly egg-shaped. Unfortunately, this doesn't make any noticeable change in what we see with our eyes. Its continually changing distance from the Earth and the changes in brightness this creates easily override any slight differences produced by its changing phase.

Life on Mars

There are so many great stories about intelligent life on Mars. Sadly, most seem to involve the Martians attacking the Earth at some point, The War of the Worlds by HG Wells is just one that springs to mind. But is this a real possibility? Could we one day have to repel an invading army from Mars?

The idea that the red planet could harbour life is not new. We have known for a long time that Mars is the most similar planet to Earth in terms of surface conditions. Early speculation about Martian life reached a peak when Percival Lowell announced in 1895 that he had (erroneously as it turns out) seen canals on the surface of Mars. According to Lowell, these canals were a final attempt by the dying Martian civilisation to bring water from the polar ice caps to their cities. In 1982 I went on a holiday to the USA and visited Flagstaff in Arizona. Lowell's observatory is situated on Mars Hill, just outside the city. During the tour of the observatory I stood outside, looking at the buildings and telescope dome, and thought about how it was here that Pluto was discovered, and Percival Lowell claimed to see canals on Mars. I had a shiver down my spine, and it wasn't from the snow falling around me.

These days we know there are no canals or interplanetary travelling Martians, but maybe at some stage in the past, there might have been life.

Our first direct attempt to look for life on Mars came with the spacecraft Viking 1 and Viking 2, which landed on the surface in 1976. No one expected their cameras to see Martians casually wandering past, so both landers scooped up some soil and did some tests. Unfortunately, their results were inconclusive and could be explained by simple, non-biological chemistry. The next possible Martian life discovery came in 1996 with the finding of what looked like microbial fossils in a meteorite from Mars found in Antarctica. Again, unfortunately, those results are still being debated. More recently, spacecraft have revealed that in the past Mars was a much wetter planet and there is evidence that even today there could still be the occasional liquid water flowing across the Martian surface.

There may or may not be life on Mars now, but the evidence suggests that at one time there could have been. The only way we will know for sure is to send a human there to look. Hopefully, we won't have to wait too long. Of course, we may not have to go at all. Maybe HG Wells was right, and they'll come to us.

The Moons of Jupiter

Jupiter is big. Larger than all the other planets combined, it deserves its title of King of the Planets. Because of its sheer size, Jupiter shines as the third brightest object in our night sky (after the Moon and Venus) despite its immense distance from the Earth.

As expected for such a big planet, it has more than its fair share of moons. Most are just small rocks, but the four largest are genuinely spectacular. All four are larger than our own moon and one of them, Ganymede, is bigger than the planet Mercury. Another, Io, is the most volcanically active body in the solar system. Callisto is one of the most heavily cratered bodies, while Europa has an ocean, potentially harbouring

Jupiter and three of its four largest moons.

life, under a moon-wide layer of ice. All four moons were discovered in 1610 when Galileo first turned his telescope towards Jupiter and as such they are sometimes referred to as the Galilean Moons. The prevailing theory at the time was that the Earth was at the centre of the universe and everything revolved around it, so it was a momentous discovery to find objects that revolved around another object other than the Earth.

Exploring for yourself: Jupiter's four largest moons

It is challenging to see the moons of Jupiter with the naked eye, but it is not impossible, as they are theoretically bright enough to be within the range of your eyes. The problem with seeing them is that Jupiter is very close, and its brilliance tends to drown out the feeble light from the moons, making it extremely difficult. I've never been able to do it, but some people claim they have seen them. If you don't have the eyes of an eagle, then try and use a pair of binoculars. The most significant problem with this method is trying to hold the binoculars steady. Some types of binoculars can be mounted on a tripod, but if a tripod is not possible, resting the binoculars against something stable, such as a railing or tree, will help hold them steady. However, if you have a telescope, use it, as a telescope will give a much better view of the moons of Jupiter than either your eyes or binoculars.

Once you have worked out how to see the moons, go out each night over a period of two weeks and observe Jupiter. On the following chart draw the moons in their relative positions and distances on the line around the central Jupiter. The next night, repeat the process on the line underneath. At the end of the two weeks, it will be possible to work out the periods for each of the moons (and hence which moon they must be) by comparing each measurement sequentially. Some nights, not all four moons will be visible as the missing moon (or moons) may be in front of, or behind, Jupiter at the time you were looking.

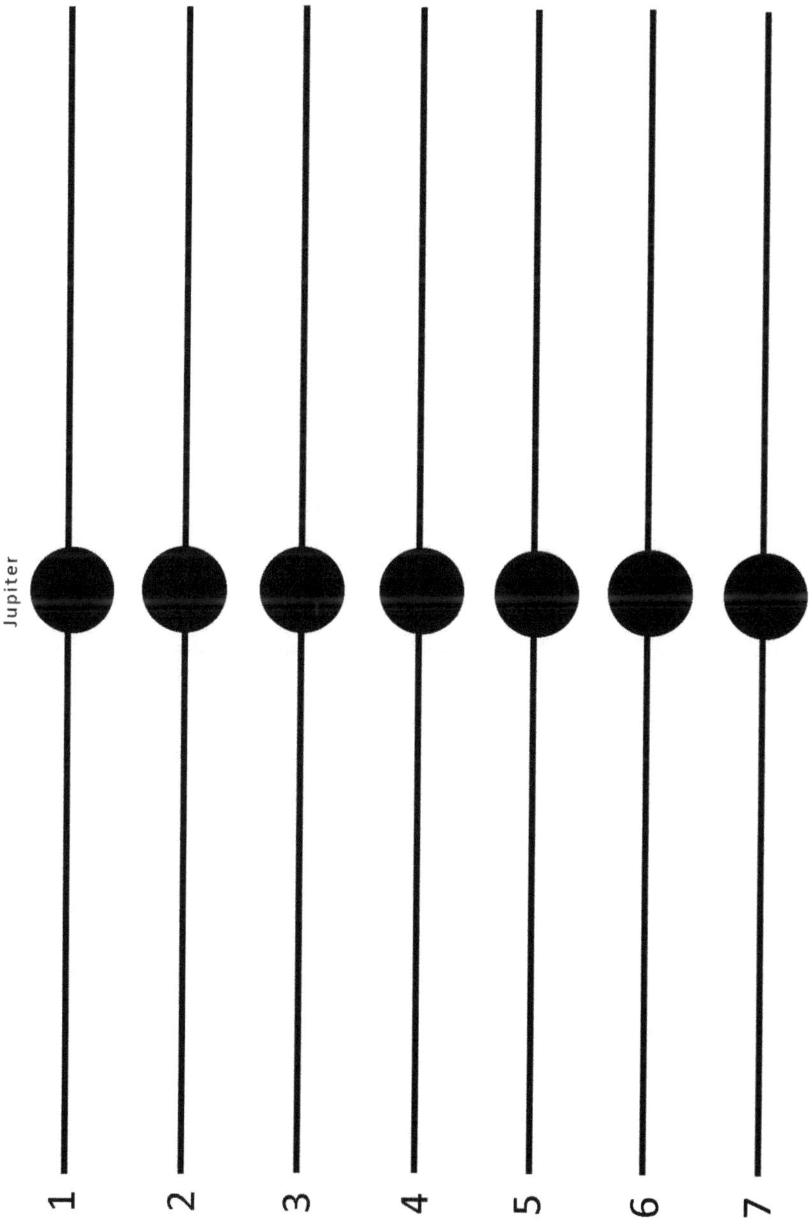

Keeping track of Jupiter's moons.

Saturn's rings

If you have used a telescope over the years to look at the beautiful planet Saturn, you may have been shocked to see its magnificent ring system occasionally disappear. It hasn't really, but through a quirk of celestial geometry, it sometimes looks as if it has.

Saturn's rings look solid from Earth, but they are made up of a countless number of small particles that range in size from about a centimetre to the size of a house. They are composed primarily of water ice, and it is the reflected sunlight off these icy particles that make Saturn's rings look so impressive and bright.

Every so often though the Sun crosses the plane of the rings. In other words, the Sun is perfectly side-on to the rings, and there will be little, or no sunlight reflected off them. If the Earth is also perfectly side-on to Saturn, the two effects combine and little, or no sunlight gets reflected to us.

Of course, nothing is ever that straightforward. There is another reason why the rings are virtually impossible to see when we are side on to Saturn. Despite their impressive appearance, there's very little material in the rings of Saturn. If compressed into a single body, it would be no more than 100 kilometres across. That means even though the ring system is vast, the material that makes them up is spread very, very, very thinly. From the inside of the ring closest to Saturn to the outside of the furthest ring is a little over 400,000 kilometres. But, on average, they are at most 100 metres thick. Viewed from a distance of over one billion kilometres, it's no wonder they disappear when we look at them side-on. On a more manageable scale, that translates to looking side-on at a 400 metre wide sheet of standard 80 gsm paper from 360 kilometres away.

Crossing the ring plane of Saturn occurs every 15 years and historically these events have given astronomers an opportunity to discover new satellites usually lost in the glare of the planet's bright ring system. Some of the brighter moons are easily visible all the time, but astronomers have discovered a number of moons during ring plane crossings. The rest have been discovered by spacecraft that either flew past or were in orbit around the mighty planet.

Saturn and its magnificent rings.

Seeing Uranus

Until relatively recently people didn't have television, computers or even electric lights. So when the sun went down, they often entertained themselves by watching the night sky. With no artificial lights to interfere with the view, each night the Milky Way was spectacular, meteors flashed across the sky, and five points of light that looked like stars slowly changed position. These wandering stars, or planets, were of particular interest. Amazingly though, despite all that watching, one planet was missed. There is a sixth planet you can occasionally see without a telescope, the planet Uranus.

An English astronomer, William Herschel, discovered Uranus in 1781 during a telescopic survey he was conducting. Uranus had been seen many times before but had been mistaken for a star, with the earliest recorded sighting in 1690 when astronomer John Flamsteed catalogued it as a star. Even though Uranus is larger than the Earth, the mistake is understandable. Since it is so far away, it just looks like a greenish star to the unaided eye. It also moves so slowly you have to watch it for decades to realize it is one of the wanderers.

Unfortunately, in modern times Uranus has become all but impossible to see using just your eyes. The planet is naturally faint, and any artificial lighting wipes it out completely.

Aphelion and perihelion

As the Earth goes around the Sun, it travels not in a perfect circle but a slightly flattened oval-shape, known as an ellipse. This means that the distance from the Sun to the Earth varies slightly throughout the year. For a few months, the Earth is a little bit closer to the Sun than on average, and for a few months, a little bit farther away.

The furthest point from the Sun in the Earth's orbit is known as aphelion and occurs in early July, right in the middle of the southern hemisphere winter. At aphelion, the Earth is about 152 million kilometres from the Sun.

The closest approach of the Earth to the Sun is known as perihelion and occurs in early January, right in the middle of the southern hemisphere summer. Generally, the Earth is about 147 million kilometres from the Sun at this point. So, during summer we are about five million kilometres closer to the Sun than at our farthest distance.

It is tempting to think that winters here in the southern hemisphere should be cooler simply because we are further from the Sun, and similarly, summers should be warmer because we are closer. However, five million kilometres isn't much compared to the average distance from the Sun of 149,597,870 kilometres, especially when you consider that aphelion and perihelion are only 2½ million kilometres either side of this average distance. Consequently, it has no significant effect on the temperature between summer and winter. Although the Earth does experience a slight shift in temperature due to this difference in distance from the Sun, it is not what causes summer to be hotter than winter. Seasonal temperatures have more to do with how concentrated or spread out the sunlight is as it strikes the Earth's surface. Additionally, the southern hemisphere is mostly water and water is slower than land to heat up and cool off. This enormous amount of water south of the equator acts as a buffer, protecting the hemisphere from big temperature swings.

The Earth's varying distance from the Sun does have one interesting effect though. The closer a planet is to the Sun, the faster it moves, and the further away it is, the slower it moves. When the Earth is at perihelion (January), it moves faster in its orbit than it does at aphelion (July). In January we travel at 109,080 km/hr while in winter we move around the Sun at a leisurely 105,480 km/hr. Looking at the length of the summer months compared to the winter months, in a typical year the half containing winter will last for approximately 186 days while the half containing summer lasts a mere 179 days. No wonder winter seems to last forever.

Barycentre

The barycentre of the solar system is continuously moving. If the term is not familiar, don't worry, the concept will be. When swinging something heavy in circles around your body it is impossible for you to turn on the same spot. The heavy weight forces you to lean back in order to balance it. In doing so, your body itself makes small circles around a point known as the common centre of mass, or barycentre. In other words, the barycentre is the point between two objects where they balance each other.

When a moon orbits a planet, or a planet orbits a star, the centres of both bodies revolve around their barycentre, a point not necessarily at the centre of the bigger body. For example, the Moon does not orbit the centre of the Earth. Both the centre of the Moon and the centre of the Earth orbit a point on the line between them approximately 4,660 kilometres from the Earth's centre (1,710 kilometres below the Earth's surface) where their respective masses balance. When the barycentre is within the more massive body, such as with the Earth and Moon, that body will appear to wobble as it moves around in its orbit.

In the same way, none of the planets orbit the Sun's centre. Depending on how big and how far from the Sun it is, each planet has its own barycentre with the Sun. If there were only the Sun and the Earth, the barycentre would be located 449 kilometres from the Sun's centre. Given that the Sun has a radius of 696,000 kilometres, this would make the Sun's wobble barely perceptible. If there were just Jupiter, it and the Sun would both orbit a point 46,000 kilometres above the Sun's surface. The Sun would then appear to wobble noticeably.

Since there is more than just one planet in the solar system, to calculate the barycentre of the solar system as a whole, you need to sum all the influences from all the planets. If they were all aligned on the same side of the Sun, the combined centre of mass would lie about 500,000 kilometres above the Sun's surface. But the planets aren't all lined up on one side. Generally, they are scattered around the Sun and as they constantly change in their orbits so too does the location of the solar system's overall barycentre. Sometimes it is located inside the Sun, other times outside. At the moment it is outside, and will remain there until about the year 2027 CE.

Incidentally, it is this visible wobbling of a star around a common centre of gravity with another object that has allowed us to discover a multitude of extrasolar planets in recent times. By observing the wobbling of the star, we can tell not just that it has a planet but also how many, how big, and how fast they orbit their star.

Planetary alignments

Some people seem obsessed with planetary alignments. Of course, some are keen amateur astronomers with an interest in seeing all the planets in one night, but others think that the planets aligning will bring on the end of the world.

We now know that the gravity of the planets causes the Solar System's barycentre to wander around, but is there any way an alignment of the planets could significantly affect the Earth?

In everyday use, when things are 'aligned' they are said to be in the same orientation, and their centres are precisely lined up. If we apply this definition to the planets, it would mean all eight planets would be hiding perfectly behind each other. Alignments between just two planets occur very rarely, so alignments between three or more planets are much rarer still. Since eclipses between planets are so rare, it does not make sense to use the word 'alignment' for them. A better term would be 'close grouping', but this isn't quite as catchy, so we are stuck with the term

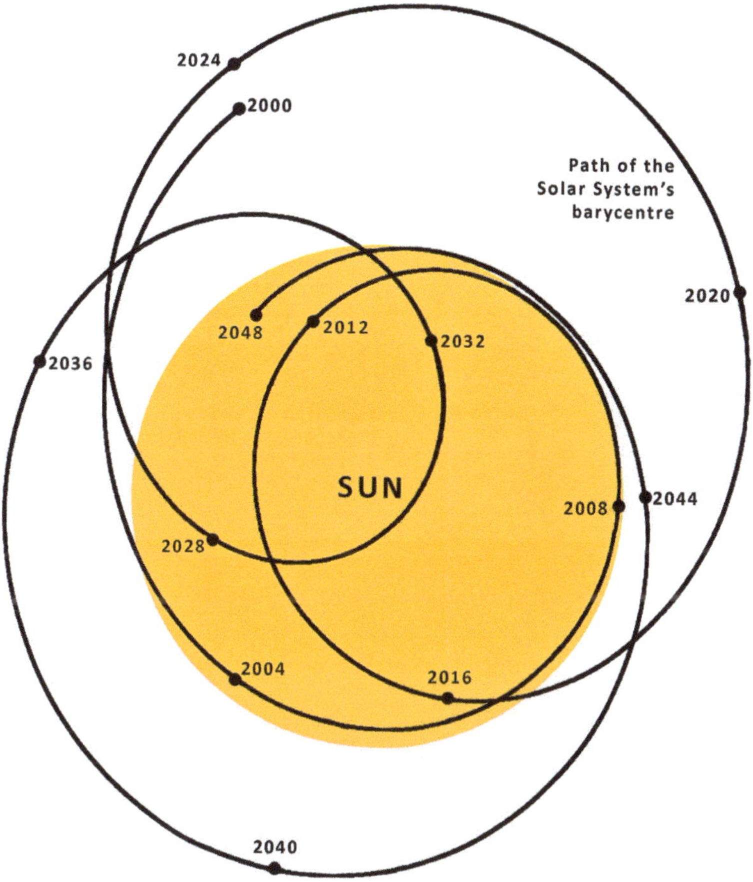

The movement of the Solar System's barycentre from 2000 CE to 2048 CE.

alignment. Momentarily forgetting that the planets don't all go around the Sun in the same plane so can't possibly line up behind each other, let's assume the planets could line up perfectly. How strong would the combined gravitational influence of the planets be on the Earth? Mighty Jupiter only pulls on the Earth about one per cent as hard as the Moon. Venus is next with only 0.6 per cent of the Moon's force. If we add them

Saturn, Venus and Mercury in 2005.

Mercury, Venus, Jupiter and Mars in 2011.

up, the total pull of all the planets combined is still only 1.7 per cent of the Moon's pull. That isn't much, but is it enough to destroy the Earth? The Moon orbits the Earth in an ellipse. That means it is closer to the Earth sometimes than at other times. At perigee, or closest approach, it is about 363,000 kilometres away and at apogee, or farthest point, it is about 405,000 kilometres away. Because of this, the Moon's gravitational effect on the Earth fluctuates by about 25 per cent each orbit. The Moon orbits the Earth in about a month, going from apogee to perigee every two weeks. So every 14 days we experience a change in gravitational effect from the Moon more than 10 times greater than all the other planets combined.

So, if gravity won't render the Earth in two, what about the tidal effect of the planets?

Since the tidal force is related to distance, the close approaches between Venus and Earth are the most critical factor in determining the combined planetary tide on Earth. The point at which the Venus tide is at a maximum corresponds to a tidal acceleration that is 1/10,000 as strong as the Sun's average tidal acceleration on Earth and over 20,000 times smaller than the tidal acceleration induced twice every day by the Moon.

The gravitational and tidal influence of the planets on the Earth is insignificant. An exact alignment of all planets (including Earth) would only produce a tidal bulge on the Sun less than a few millimetres in height, in essence, zero. But this does not stop some people from believing the planets can influence events on the Earth, even to the extreme of destroying all life on the planet. Their beliefs seem to be bolstered by the conjunction of alignments with important calendrical dates, such as the 2012 Mayan association. Other famous 'end of the world due to planetary alignment' dates occurred in 1982 and the year 2000. As far as I can tell, the world didn't end on any of those dates either.

Since they never happen, let's forget about exact alignments and talk about close groupings of the planets.

For the planets Mercury through to Saturn, as viewed from Earth, in the period between 3100 BCE and 2750 CE the last relatively close alignment was in May 2000, where they were within 20 degrees of each other. I remember looking at it and being suitably impressive at the time. Within this almost 6,000 year period, however, 55 alignments are at least as close as the one of May 2000.

On average, close alignments occur about nine times every 1,000 years. Before 2000 CE, the last alignment that was as close occurred in February 1962 CE. If we look at the closest alignment during this period, it happened in late February 1953 BCE when they were just four degrees apart. That would have been a spectacular sight. The next alignment of note occurs around 8 September 2040 CE when they will be eight degrees apart. That is not as wide as your closed fist held at arm's length. Make sure you put this date in your diary as you will not want to miss it.

Mercury is moderately hard to see from the Earth, so what if we take Mercury out of the list and look at the remaining planets you can see with the unaided eye, that is, Venus through to Saturn. The closest alignment in the period stated above will be in February 2378 CE with a width of just three degrees. Incidentally, the last big one was in June 1564 CE at four degrees separation.

If we extend the list to include the planets we can't see with the unaided eye, then Mercury through to Neptune have their next significant alignment in March 2673 CE at 25 degrees. The most recent event occurred in January 1665 CE, but at 28 degrees it was not quite as tight a grouping.

The end of the solar system

Five or six billion years from now, the Sun will deplete the hydrogen in its core and begin an ever-quickening race toward its ultimate fate as a white dwarf, a stellar corpse no bigger than the Earth.

However, before it gets to that stage, as it nears the end of its life the Sun's core will contract and grow hotter while its outer layers expand and cool, turning the Sun into a red giant star. It will expand to many times its current size, swallowing Mercury, Venus and probably Earth in the process.

As the red giant Sun expands, the gravity holding its outer layers will relax. Pushed by the radiation still being produced in its core, the gas in the outer layers will stream into space, forming a hot, expanding cloud known as a planetary nebula. The energy still being produced by the core will illuminate the cloud until the gas making up the nebula disperses and disappears from view. As the gas moves outward into space, it will sow the seeds of new stars and planets throughout our celestial neighbourhood. Most of the carbon (and nitrogen, oxygen, and other heavy elements) in the universe probably formed in the hearts of red-giant stars before being hurled into space as those stars died. New generations of stars formed from the debris of these older stars then contain more carbon than their stellar ancestors.

So, when our Sun dies it will not only put on a beautiful display, it will scatter the raw material for new stars, new planets and perhaps new life throughout our corner of the Milky Way galaxy.

Retrograde

Like every object in the heavens, during a single evening a planet will move from east to west across the sky due to the Earth's daily rotation. But if you pay close attention, you will notice that slowly but surely from night to night the stars also appear at the same time to have moved a little bit further to the west. This is due to the Earth's rotation not being precisely 24 hours. Concentrating on the planets, however, shows that compared to the stars, the planets slowly drift east each night.

Every so often a planet will appear to halt its slow eastward movement and remain stationary compared to the stars. It will then slowly start moving westward each evening. It doesn't move much each night at this time around its stationary point, but as the weeks go by its westerly march becomes much more apparent. Eventually, it will once again stop its westerly motion compared to the stars and resume its original easterly drift. The apparent change in a planet's direction is known as retrograde motion. All of the planets further out than the Earth do it and, as bizarre as it may seem, it is a natural consequence of us living on another planet which is also in motion around the Sun.

For ancient astronomers, this retrograde motion of the planets was hard to explain. For centuries the best model had the Earth at the centre of the universe and the model had difficulty explaining the observed retrograde motion. A complicated system of planets moving around the Earth in small circular paths (called epicycles) that in turn moved on larger circular orbits was required. At best, it was cumbersome and not entirely accurate.

Eventually, Copernicus proposed a more straightforward system that ultimately proved to be correct. His model explained retrograde motion by having the planets circle the Sun, with the planets further from the Sun moving slower than those closer in. Retrograde motion can then be explained simply as our view from the surface of a faster-moving planet as it catches up to the slower moving outside planets and passes them by.

Artificial satellites

Since the early 1960s, a lot of satellites have been put into orbit. The largest is only a few metres across, but all of them are metallic and act as giant mirrors reflecting the Sun's light down to the ground. Looking like bright, star-like points moving at a constant speed across the sky, on any one night there are dozens visible.

They move in all directions, with some regularly fading and brightening as they tumble in their orbit. Because most satellites are in a low Earth orbit, they also travel quite quickly, taking only an hour or two to go around the Earth. As you watch a satellite moving towards the east, it can suddenly disappear, even though it hasn't gone below the horizon. This is because it has run into the Earth's shadow and no longer reflects the Sun's light.

In recent years, more substantial satellites have been placed in orbit about the Earth. If you know how to pick them from the rest, you can easily see the International Space Station and the Hubble Space Telescope. The James Webb Telescope will add another sizeable shiny object that will also be visible from the ground. More recently there have been trains of satellites placed into orbit. These appear as strings of bright lights moving across the sky. No doubt someone has already reported them as a UFO invasion fleet.

Generally, the best time for satellite spotting is in the 90 minutes immediately after sunset or before sunrise. While it is dark here on the ground at this time, up where the satellites are the Sun is still shining, and we can see the bright object against a dark sky. For the rest of the night the satellites hide in the Earth's shadow and cannot be seen.

Meteors

As comets travel around the Sun, they leave behind a dusty trail of debris. If this trail happens to cross the Earth's orbit, when we pass through it we get a spectacular display known as a meteor shower. The particles of dust, on average about the size of a pinhead, collide with our atmosphere at phenomenal speed, producing a short streak of light commonly known as a 'shooting star' or 'falling star', a bad name as they are of course not actually stars.

The usual explanation for what produces the bright streaks of light is that it is friction with the air that causes the meteor to melt, but there is very little friction between the meteor and the air. The hot, compressed air in front of the meteor stays far enough in front to create a buffer of relatively slow-moving air directly in contact with it. So, if it isn't friction that melts the meteor, what is it? When the bit of dust enters the upper atmosphere, it compresses the air in front, and when a gas compresses, it heats up. The air is compressed so much and gets so hot it melts the side of the meteor facing this furnace of heated air. As the meteor melts, it releases different chemicals, and these emit a bright light. The meteor itself can also glow as its surface melts. All factors combined create the meteor we see flashing across the sky.

The bright streak of a meteor happens very high in the atmosphere, at altitudes around tens of kilometres. As the meteor interacts with the atmosphere it quickly slows down, and once it has slowed to below the speed of sound, the air in front is no longer compressed quite so much, and the meteor stops glowing. Regular friction then takes over and slows the meteor even more. It then takes just a few minutes to pass the rest of the way through the atmosphere. Larger pieces around the size of a grain of sand or small pebble cause the more spectacular meteors, sometimes seen breaking up during their fiery encounter with our protective atmosphere. These brighter meteors are known as fireballs. If it manages to survive and hit the ground, we then call it a meteorite.

About 40,000 tonnes of material from space hit us every year. Given that water covers 70 per cent of the Earth, we can expect only 12,000 tonnes of that to fall on land. Most of it is in the form of dust raining down, although there are some larger pieces from asteroids and comets. For example, about five per cent of all meteorites come from the asteroid Vesta, approximately 400 fragments of the Moon have been found, and 130 pieces of Mars have fallen onto the Earth's surface. Think about those numbers for a moment. I don't know about you, but I find them amazing.

One of the most famous meteorite sightings in modern history, the Peekskill meteorite, broke up over the United States in 1992. At its peak brightness, it was comparable to the Full Moon, and during its 40 second long flight, the fireball covered approximately 800 kilometres before it finally came to the ground and attacked a perfectly innocent parked car.

The only known meteor to have killed humans, albeit indirectly, is the Tunguska event. Occurring in Siberia in 1908, the explosion of an asteroid 10 kilometres above the ground flattened an estimated 80 million trees over an area of 2,000 square kilometres. There have been other non-fatal human encounters. In 1927 a bean-sized stone was found resting on the head of a five year old girl in Japan. A young boy in Uganda claimed that he had been hit by a small meteorite while standing under a banana tree, and a German boy claimed he too was hit by a meteorite while walking to school. None of these caused injury, but there have been some that did. The first known case of someone injured by a space rock occurred in 1954 when a four kilogram meteorite crashed through a roof and hit a lady while she was sitting on her lounge. She was badly bruised but otherwise unharmed. In a more indirect instance, a meteor exploded over Chelyabinsk in Russia in 2013, causing widespread damage and injury.

Impacts large enough to create big craters and do significant damage are estimated to occur on average every 1,000 years. Every few tens of millions of years, we have impacts significant enough to cause mass extinctions that threaten the survival of all advanced life forms on Earth. So the next time you see a shooting star keep in mind that, although extremely rare, occasionally they do hit something. The prospect of being hit on the head by a rock from space is extraordinarily unlikely, but then it is worth noting that it was a meteorite that did away with the dinosaurs and another that created the Moon when it collided with the Earth.

Exploring for yourself: Meteor spotting

Lie outside on your back for a few hours and look up. On any clear night you will see, on average, five or six sporadic meteors per hour coming from all over the sky. However, at regular times each year the Earth passes through streams of dust left behind by comets, and when it does, something known as a meteor shower is visible. The best showers can have up to 80 meteors streak across the sky each hour.

Even during a shower that has 80 meteors per hour, you still have to be patient, but if you spend more than a few minutes outside you will be treated to a great show. At these times, due to perspective, the meteors all seem to come from the same spot in the sky, known as the radiant. Meteor showers are best visible during the early morning, as that's when the Earth is spinning into the direction it travels around its orbit and consequently runs into more bits of dust. Of an evening, the Earth is spinning away from its motion around the Sun so only the fastest specks of dust can catch up to us, which is why we only see a few sporadic meteors of an evening.

On infrequent occasions, something known as a meteor storm can occur. The last one was in 1966 when there were an estimated 144 thousand meteors per hour. That's 40 meteors per second. By any definition that would have been spectacular to see. Interestingly, the book *The Day of the Triffids* was written by John Wyndham in 1951, but when you read his description of the meteor shower that blinded everyone you can't help wondering if he had a premonition. His description almost perfectly matches that given by eyewitnesses of the 1966 storm 15 years later.

Meteors come in different brightnesses, length of duration, and more rarely, colour. They can have a long train behind them or periodically flare. Given the unpredictable nature of individual meteors, the best way to observe them is to lie outside for a while and enjoy the view. The only things required are your eyes and perhaps warm clothing or a blanket.

Close encounters

Near Earth Objects, such as an asteroid passing the Earth less than twice the distance of the Moon away, may not be deemed newsworthy, but in astronomical terms, it is way too close for comfort. In recent years it seems more and more of these objects are barely missing the Earth. That isn't entirely true as they have always been passing by, it's just that we are discovering more of them. In the past, they passed by without being noticed.

So why are we finding more of these nearby rocks in space? Following a few near misses that were discovered long after they had gone past the Earth, astronomers and governments around the world realised that the potential for unannounced disaster from an asteroid impact was much greater than previously suspected. That prompted a survey to locate at least 90 per cent of the large near-Earth asteroids. Having reached that target, we are now well on the way to finding almost all of them.

Knowing where the near-Earth asteroids can be found is a worthwhile endeavour in itself, but even more important is discovering any that could potentially hit us. The International Astronomical Union realised this and established an international foundation to locate what are called Potentially Hazardous Asteroids (PHAs). These are not just the large ones, but the smaller asteroids as well. A rock from space doesn't have to be huge to potentially cause immense damage.

Discovering small rocks in the emptiness of space is not easy. Currently, the best way of finding asteroids is by repeatedly surveying large areas of the sky with telescopes. By looking at the same patch of sky over and over again, hopefully any asteroids will show up as faint points of light that have changed position. Once a possible candidate has been discovered, it can then be tracked over time to determine its orbit and assess whether it is a potential threat to the Earth. The system isn't perfect, and a few have been missed, but it has also picked up a few that subsequently crashed into the Earth. Knowing where a PHA came from and following its path through the atmosphere allows astronomers to more accurately predict the trajectories of other PHAs.

Asteroid names

What do sunflowers, volcanoes, musicians, actors, the United Nations and the lost island of Atlantis have in common? They all have asteroids named after them.

Asteroids, or minor planets as they are sometimes known, have names assigned to them after a long process taking years. It begins with the discovery of an object that cannot be matched with any known object. It is then given a provisional designation according to a formula and the date of discovery. When the orbit is determined well enough to predict the asteroid's position far into the future reliably, it receives a permanent number designation, such as 6,723 or 317,992. The discoverer is then invited to suggest a name for it. They have 10 years in which to come up with something, but I would have thought if you can't think of a name within one year you are never going to think of one. The discoverer is given 10 nonetheless.

Proposed names are judged by an international group of astronomers and must conform to set rules. Names should be no more than 16 characters in length, preferably one word, pronounceable, unoffensive and not too similar to an existing name. Names of individuals or political and military events are unsuitable unless 100 years have passed since the death of the individual or occurrence of the event. Pet names are also discouraged, and names of a commercial nature are not allowed.

There are now well over 500,000 known asteroids, and the number grows every year. Not all of these are named, and not all of them ever will be named. Only a small percentage have one, while the rest retain their number. When you look at the list of asteroid names, the origin of some of them are instantly recognisable. Many, however, are obscure and related to the discoverer in some unknown manner. As I've already mentioned, names are not just restricted to people. Animals, geological features, countries, fictional characters and even institutions all get a go.

Eclipse

A common misconception is that we see the Moon and planets because they give out light. Since the Sun is the only body in the solar system that produces its own light, this is incorrect. That means all other objects, such as the Moon and planets, can only be seen because they reflect sunlight. Now, and this may seem obvious, an often overlooked consequence of the Sun as the only source of light is that everything must, therefore, cast a shadow. Usually, these shadows are invisible in the blackness of space, but every so often they can be seen, such as when the Earth and Moon pass through each other's shadows.

When the Moon lies directly between the Earth and Sun, its shadow falls onto the Earth. If you happen to lie under this shadow, the Moon completely blocks out the Sun and it will, of course, go dark. This is one of the most spectacular astronomical events you can witness and is called a total solar eclipse. The Moon's shadow doesn't completely cover the Earth, just a small patch that races across the Earth's surface as the Moon continues to move. Because of this, the eclipse only lasts for a few minutes before the shadow moves on from your location.

Sometimes the Moon only passes partly in front of the Sun, and we get to see the edge of the Moon's shadow. When this happens, we have a partial solar eclipse.

Through a strange celestial quirk, the Moon and Sun appear about the same size in the sky, which is why it is possible to get total solar eclipses. They aren't the same size, the Sun is about 400 times larger than the Moon, but it's also about 400 times further away, so size and distance balance out and they appear to be the same size.

Because of their elliptical orbits, sometimes the Moon is closer to the Earth than usual and sometimes the Earth is closer to the Sun than usual. If the Earth is at its closest to the Sun, the Sun will appear slightly larger. If the Moon is at its furthest point from the Earth, it appears somewhat smaller. If either of these conditions occur close to each other at the time of a solar eclipse, the Moon then doesn't appear big enough to cover the disc of the Sun completely, and we get something known as an annular eclipse. If you can safely look at the Sun at this time, you would see the Moon's dark disc surrounded by a bright ring of the Sun's surface. Hence the name annular eclipse.

The problem with looking directly at solar eclipses is that it means you are looking directly at the Sun. The Sun is far too bright for your eyes to handle and doing so can cause severe and permanent eye damage, if not blindness. An excellent general rule is NEVER look directly at the Sun, but there are safe ways to observe solar eclipses. Special filters for looking at the Sun are available for telescopes, but these tend to be expensive and not something you are likely to have unless you are seriously into solar viewing. A more convenient and cheaper way of observing the Sun is by projecting the Sun's image onto a screen and then looking at the image.

Now, what happens when the Moon passes directly behind the Earth and encounters the Earth's shadow? Passing completely through the shadow means the Moon no longer reflects sunlight and it goes dark, producing a total lunar eclipse. These tend to be more commonly seen than solar eclipses because they are visible to anyone on the side of the Earth facing the Moon at the time. When the Moon only partly passes into the Earth's shadow, only part of the Moon's disc goes dark, and it is called a partial lunar eclipse. Unlike solar eclipses, lunar eclipses pose no problem to look at so you can use just your eyes. Like any shadow, the Earth's shadow also a lighter outside section called the penumbra. The casual

A partial solar eclipse.

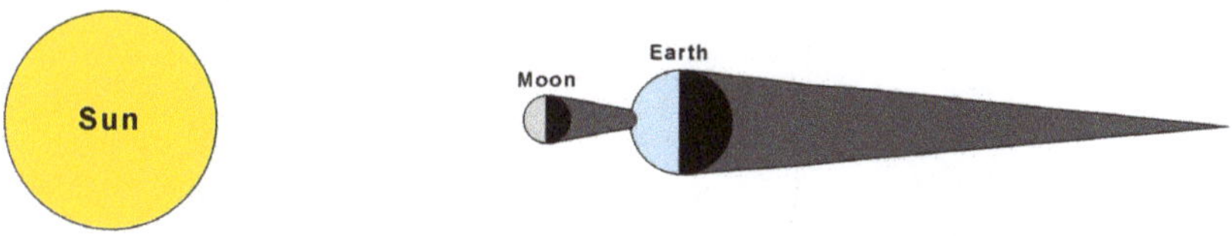

Solar eclipse.

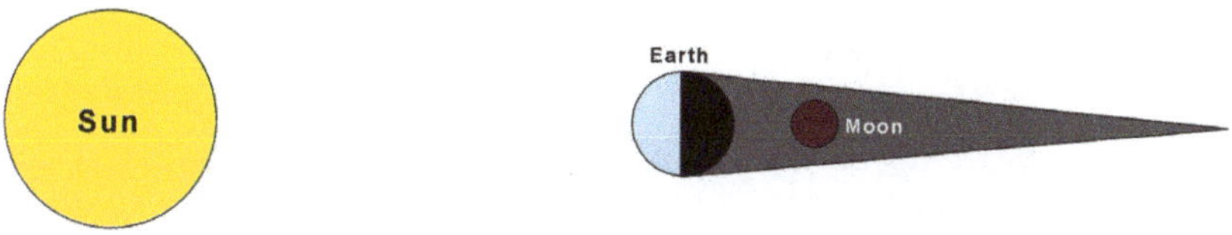

Lunar eclipse.

A partial lunar eclipse.

A total lunar eclipse with the typical red colour caused by refracted light from around the Earth.

observer, however, is unlikely to notice when the Moon passes through the Earth's penumbra.

Between 3000 BCE to 3000 CE, the Earth experiences 14,255 solar eclipses (partial and total) and 9,215 lunar eclipses (partial and total). On average, this works out to just over two solar eclipses and about one and a half lunar eclipses each year. Of course, eclipses don't happen like clockwork, and the intervals between eclipses vary, so any one year can have between two and five solar eclipses and between zero and three lunar eclipses. Typically, there is a total solar eclipse somewhere on the Earth's surface every one to two years. But since a total eclipse is only visible from a very narrow path, it is rare to see one from any single location, and you usually have to wait about 375 years to see a second total solar eclipse from where you live.

Transits

Back in 2004 CE and 2012 CE, a rare and interesting astronomical event took place. For the first time since 1882 CE, the planet Venus passed directly between the Earth and Sun, and we witnessed what is known as a transit of Venus. Unfortunately, we will now have to wait until the year 2117 CE before we see Venus do it again.

Venus, however, is not the only planet to put on a show with the Sun, as Mercury also transits the solar disc. Thankfully a lot more often than Venus does. Looking from here on Earth, Mercury and Venus are the only planets that ever pass directly between the Sun and us and are consequently the only ones we ever see pass over the Sun's disc. Transits don't occur every time Mercury and Venus move between the Sun and the Earth because their orbits are tilted slightly to ours, and most of the time they appear to pass above or below the solar disc. But fortunately, the Sun appears quite big to us, so the alignment doesn't have to be perfect and consequently transits of either planet aren't that rare.

On average there are 13 Mercury transits each century, roughly one every 7½ years, while Venus transits the Sun only 81 times in 6,000 years, twice eight years apart and then not for over a century. Although you could within your lifetime expect to witness several individual transits, having both planets transit in the same year is extremely rare. It occurred back in

1631 CE and 1769 CE and the next time won't be until 2611 CE.

An even rarer event is to have transits of both Mercury and Venus occurring at the same time. No one knows when the last one happened, but a date to put in your calendar for the next one is 26 July in the year 69,163 CE. The next transit of Mercury during a partial solar eclipse is not quite so long a wait and occurs on 5 July 6,757 CE. If you want to see Venus transit during a solar eclipse, you will have to wait until 5 April 15,232 CE. Since the chances of still being around to see any of these events are slim, anyone currently alive wishing to observe a transit will have to be content with a simple transit of Mercury.

These days we no longer use the transit of Venus for any significant scientific purposes, but back in the 18th and 19th centuries astronomers used them to establish one of the largest and most meaningful measurements, the scale of the solar system. They are of particular interest to Australians since it was Captain James Cook's voyage to Tahiti to observe the 1769 transit of Venus that led to the European settlement of the Australian continent.

Incidentally, the telescope bought to observe the 1874 transit of Venus from Australia is still in operation at Sydney Observatory. It may be the only telescope left in the world that will have been used to observe three transits of Venus.

Planetary transits

Mercury and Venus are the only planets that ever pass directly between the Sun and us, meaning they are the only planets we get to see pass over the Sun's disc from here on the Earth. There is also the ever so remote chance that we also get to see Venus transit across Mercury. But given Venus appears much bigger than Mercury, it would technically be an occultation. Consequently, we will limit our discussion to viewing planets passing across the face of the Sun.

As hinted, transits can be seen from other planets too. They are not purely an earthly phenomenon. Besides Mercury, the closest planet to the Sun, every planet will at some stage have another one pass directly between it and the Sun, resulting in a transit. Naturally, the further out the planet is, the smaller the Sun will appear, and the more perfect the alignment has to be. This means the further out in the solar system you live, the less likely you are to see a transit. The rarest transit of all is that of Uranus crossing the Sun as seen from Neptune. Between the years 0 CE and 4000 CE it does not happen. Not even once. Never.

Humans are not likely to be standing on the surface of Neptune, Uranus, Jupiter or Saturn in the foreseeable future. But they may be on Mars in a matter of decades. So what about the next transits as seen from Mars? Transits of Mars' small moons across the Sun have already been seen by rovers on the surface of the red planet, but in the year 2030 CE Venus will cross the Sun's disc and in 2084 CE the home planet, Earth, will be seen directly in front of the Sun. Will any Earthling be there to see this event? I hope so, as that would be a sight not to be missed.

CHAPTER 16

Musings of a stellar nature

—

Why do stars twinkle?

If you ever look at the stars you quickly realise that they all seem to twinkle. The reason why is not with the stars themselves, but how we get to look at them. We stand on the solid surface of our planet surrounded by a thin layer that most of the time we tend not to think about, the atmosphere. Usually, the atmosphere is invisible to us, but when we try to look at a star, it becomes apparent just how hazy and turbulent the atmosphere really is.

Since all stars (except for the Sun) are so far away, when their light finally reaches us it is nothing more than a pinpoint. As this tiny amount of light enters the atmosphere, it gets jiggled about so much and at such a rapid rate by the air that when we get to see it, the pinpoint appears to be twinkling. Astronomers refer to this twinkling due to the atmosphere as 'seeing'. Since the steadiness of the atmosphere is continually changing, the amount of twinkling also constantly changes. Some nights the stars appear steady, while on other nights they twinkle relentlessly.

Is it possible to fix this problem of unsteady seeing conditions? Unfortunately, to stop the twinkling completely, we would have to get rid of all the air. Since this would make breathing difficult, there are better and easier solutions to the problem. One solution is to get entirely above the atmosphere. This is why telescopes have been put into space. Another solution is to look through as little air as possible. Some of the world's biggest telescopes are found on top of extremely high mountains for precisely this reason. These large telescopes also employ some fancy engineering known as adaptive optics to cancel out as much of the twinkling as possible. But is there a solution to the problem of twinkling for someone who doesn't have either spacesuit or mountain? Yes. Look straight up. That way you are looking through the minimum amount of atmosphere and the stars will generally twinkle much less.

One final thing. Many people falsely believe that stars twinkle and planets don't. This is not correct. Since more light reaches us from a planet than from a star, planets overhead may not twinkle quite as much as the stars do, but they do still twinkle, especially when they are low on the horizon.

Star brightness

When you look at the night sky, one of the first things you notice about the stars is so evident that no one ever thinks about it, they are all different brightnesses.

But why do some stars appear bright and some faint? The first and primary reason is that stars naturally differ in brightness. Some stars shine exceptionally bright while others barely raise a dim glow. For example, found within the constellation of Orion is the star Rigel, one of the brightest nearby stars and easily visible from most places on Earth. It shines over 40,000 times brighter than the Sun.

The second reason why stars differ in brightness is distance. The Sun, our closest star, appears the brightest and largest, but in comparison to most of the stars visible to our eyes, it is at best an average star. So, even though Rigel is immensely brighter than the Sun, it doesn't appear so because of its greater distance from us. The closest star apart from the Sun is Alpha Centauri (the pointer furthest from the Southern Cross), and even though it is the nearest, it is still approximately 278,000 times further away than the Sun. No wonder the stars look like faint points of light.

The third reason for the variation in the brightness of stars is size. Stars come in all different sizes, and this also contributes to how bright they appear. Visible also in Orion is the massive star Betelguese. At around 650 times the diameter of the Sun, it truly is a big star, and because of this, it shines brightly in our sky. At the other end of the scale is the star 2M0939, about 30 times the mass of the planet Jupiter and found in the constellation of Antlia. Due mainly to its size it shines approximately 1,000,000 times fainter than the Sun and is impossible to see with anything but the largest telescopes.

It is a combination of all three factors - inherent luminosity, distance, and size - that produce the different brightnesses we see in the stars. Perhaps more importantly, it's this variety that enhances the majestic display we see when we look up into the wonderful night sky.

Exploring for yourself: How many individual stars are visible from where you live?

It is impossible to look at the night sky and count how many individual stars you can see. There is, however, an easy way to estimate how many are visible. As a rough approximation, we can assume the stars are evenly spread across the entire sky. The number of individual stars can then be estimated by taking several small samples randomly located about the sky and counting the stars in each of those samples. The total number can then be calculated using the individual tallies. The only place where the stars are not evenly spread is in the Milky Way, but within the Milky Way there are so many stars that they blend together and cannot be differentiated into individual stars anyway.

To have a useful representative sample, you need to know the area of the sky your sample covers. To give a set size for the sample area would also entail having to provide a set distance to hold it from your eyes. The problem with this method is that it is very fiddly. An easier way is to work out how big your sample area has to be if you held it at arm's length. It will be a different size for each person, but it turns out that for everyone when they hold their stretched out hand at arm's length, it measures about 15 degrees across the sky from thumb to little finger. Generally, if you have long arms, you also have a bigger hand, and a smaller hand means you have shorter arms, so it all evens out. It is not a precise measure, but it's good enough for our purposes.

Measure how far apart you can stretch your hand and then in a piece of cardboard (or anything equally suitable) cut out a square with this measured distance for each side. When you go outside and hold it at arm's length, it will frame an area in the sky about 15 degrees by 15 degrees, or 225 square degrees. The sky has a total area of about 41,253 square degrees, so when you look through your template, you are looking at about half a per cent of the entire sky area. As an example, my square is 23 centimetres on each side.

Hold the square up to a part of the sky and count the individual stars you can see through it. Write the number down so you don't forget. Repeat this 10 times in 10 different parts of the sky. The closer a star is to the horizon the more it will dim due to something called 'atmospheric extinction', which basically means you are looking through more of the

A template for counting stars.

atmosphere and this causes light from the star to be lost. So, try and make your test areas all higher than 20 degrees above the horizon. That way you won't lose too many stars due to the atmosphere. Once you have the 10 numbers, add them together to get how many stars you have counted in a total of five per cent of the sky. Multiply by 20 and you will have a rough estimate of how many individual stars you can see from your location. The number should be around 6,000 stars, although given the rough calculations used, potential light pollution, and whether the Moon was visible at the time, there may be some degree of variation. It should at least be in the ballpark. When I did it from my backyard in the middle of a light-polluted country town of 40,000 people, my numbers totalled 207. Multiplied by 20 gives 4,140 stars. Since the number of stars visible from a dark location is around 6,000, it shows how much light pollution has affected the view from my home.

Star colours

If you spend a night looking at the stars with just your eyes, you would probably say they were all plain white. Stars, however, are not only white, they come in a range of colours. The reason we see them as a bland, uniform white colour is because they are so far away that the amount of light that eventually reaches us is too little for our eyes to register any colour. We would need to be a lot closer to see the brilliant colours they are.

Stars are different colours depending on their temperature. If a star is relatively cool then the peak colour it emits is red and the star will appear red. If the star is relatively hot, the primary colour is blue. When the temperature is in between, and the primary colour it emits is green, instead of a green star, we see a white star. Although there is more green than any other colour, there is a sufficient amount of all the different colours present to blend together and make it look white.

Most stars appear white to the unaided eye because of their faintness, but there are a few stars that do show some hint of colour. Antares (the heart of Scorpius), Aldebaran (the eye of Taurus) and Betelgeuse (the shoulder of Orion) are some of the few visibly red stars. In the constellation of Bootes (the herdsman) is the bright, orange looking star, Arcturus. As a pointer to the Southern Cross, Alpha Centauri is easy to find and yellow in appearance. And Spica, a distinctly bluish coloured star, makes up one of the hands of Virgo.

One final star you might like to find is the source of a lot of debate. Many people claim it has a distinct green tinge, despite green being an impossible colour for a star to have. Located not quite two hand spans at arms-length to the west of Antares are two moderately bright stars that make up the constellation Libra. Have a look for yourself at the northern one, known as Zubeneschamali, and see what you think.

Antares

Antares, the brightest star in the constellation Scorpius, gets its name from looking similar to the planet Mars. Whenever Mars passes close to the star, you can see that they are very close in brightness and colour. Since the Greek name for Mars is Ares, and because of its similarity to the planet, the star was called Anti-Ares, or Antares, meaning 'a rival of Ares'.

Like Mars, Antares has a distinct reddish hue, and because of this it represents the heart of the scorpion. It is one of only four stars in the night sky that visibly appear reddish. But unlike Mars, the red colour comes not from being rusty, but from its temperature. As a star reaches the end of its life, it expands, and as it expands, it cools down and becomes red in colour.

The statistics of Antares are spectacular. It is the second largest star visible to the unaided eye. If you were to replace the Sun with Antares, it would engulf the entire inner solar system to out past the asteroid belt. That makes it over one billion kilometres across or almost 800 times the diameter of the Sun. Antares is also 10,000 times brighter than the Sun, but due to its distance of 550 light-years, it only appears as the 15th most brilliant star in the night sky.

Betelgeuse

Another visibly red star is on the opposite side of the celestial sphere to Antares. Find the saucepan in the summer evening sky and not far below you will see the reddish coloured star called Betelgeuse, one of the most interesting stars in the sky.

Different star colours in the constellation Orion (the hunter).

The constellation Scorpius (the scorpion) with the orange coloured star, Antares.

If you are wondering how to pronounce the name Betelgeuse it seems there is no real consensus, but you could do a lot worse than saying 'Beetlejuice'. Most people will know which star you are talking about if you do. Like Antares, Betelgeuse is magnificent by any definition. Only 10 million years old (compared to the Sun's age of 4½ billion years) if it were side by side with our Sun it would be 10,000 times brighter in visible light, even though its temperature is only 2,800 degrees and its mass only 12 times that of the Sun. It appears as the 10th brightest star in the night sky, but if we could perceive all wavelengths of light it emits it would easily be the brightest star.

Betelgeuse is one of the largest stars known, but determining its size is problematic. Measuring its distance is also difficult because the parallax angle is small compared to its angular size. Observations even indicate that Betelgeuse has shrunk by more than a distance equal to the orbit of Venus since 1993, and no one knows why.

Betelgeuse is most likely currently burning helium in its core with a surrounding shell consuming hydrogen. Someday it will eventually run out of fuel and explode in a spectacular supernova explosion, but it is just as likely to explode millions of years from now as it is tomorrow. At a distance of roughly 600 light years, when Betelgeuse does blow up it is thankfully too far away for the supernova to harm us. However, we will see an amazing sight, a very, very bright star that will brighten enormously for a few weeks until it can be seen during the daytime, perhaps even becoming as bright as the full Moon.

At the end of 2019 CE and the beginning of 2020 CE, Betelgeuse dramatically changed its brightness. Some people speculated that this might be a sign the end is near for the star. But after a few months of plunging to its faintest on record, Betelgeuse started to recover and brighten again. Astronomers are using this unusual dimming episode to study the star in unprecedented detail.

Incidentally, Betelgeuse is moving away from Orion's Belt (where it is thought to have formed) at 30 kilometres per second (108,000 kilometres per hour). Not excessively fast, but not dawdling either.

Aldebaran and the Hyades

Looking like a 'V' shape, the Hyades is a cluster of stars close to the belt of Orion. The stars within this distinctive group are all related to each other, having formed from the same cloud of gas and dust around 625 million years ago. Representing the head of Taurus, the bull, the Hyades lie just 151 light-years away, making them the closest cluster of stars to the solar system.

The fainter 'eye' of the bull's head is a star known as Epsilon Tauri. In 2007 a planet about seven times as massive as Jupiter was discovered orbiting this star roughly once every 1½ years. The other eye, known as Aldebaran, is the one star in the V shape entirely unrelated to the rest, despite its fantastic fit as an eye in the head of the bull. Located only halfway to the Hyades, Aldebaran was once even closer than its current distance of 65 light-years. About 320,000 years ago, it was just 21½ light-years distant. At that distance it would have been the brightest star in the night sky, a distinction it would have held for some 200,000 years. At Aldebaran's current distance it now only shines as the 13th brightest star in the night sky, even though it is still about 150 times the Sun's luminosity. As the changing distance suggests, Aldebaran is moving away from the Sun at a speed of 54 kilometres per second, making it one of the fastest receding stars we can see with the unaided eye.

Appearing slightly reddish to the eye, Aldebaran is a massive orange coloured red giant star some 44 times the diameter of the Sun (about 61 million kilometres). It takes nearly two years to spin around once, making it also one of the slowest rotating stars.

In about two million years the Pioneer 10 spacecraft, which flew past Jupiter in 1973, will also fly past Aldebaran.

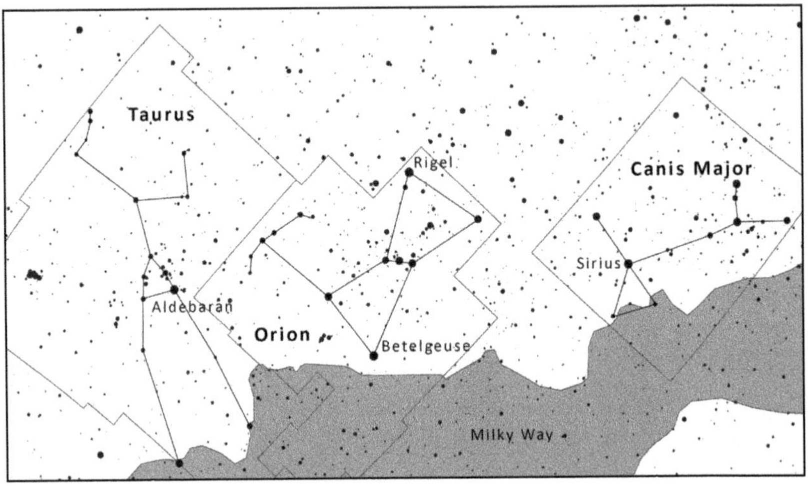

The constellations Canis Major, Orion and Taurus.

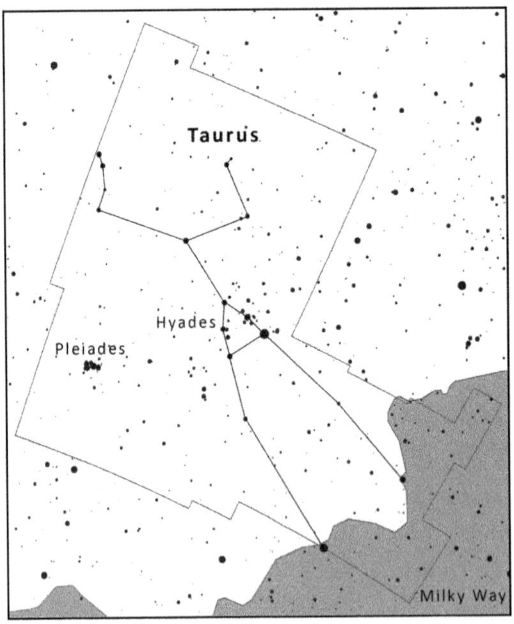

The constellation Taurus (the bull).

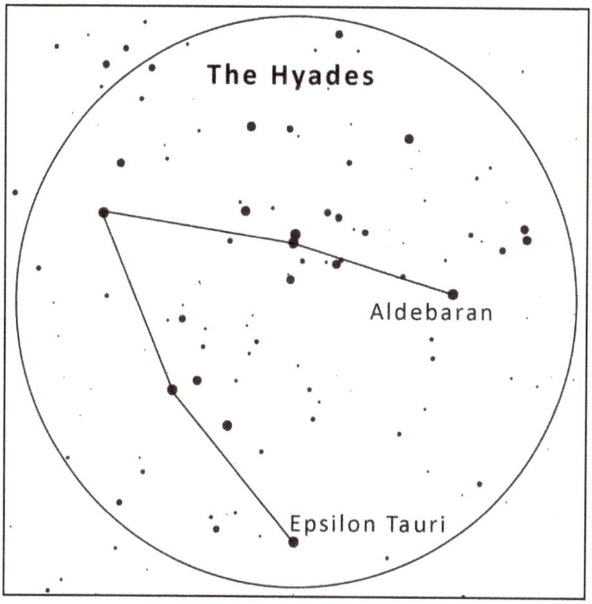

The Hyades.

The Pleiades.

The Pleiades

Still within Taurus, the Pleiades is perhaps the most well-known cluster of stars in the sky. Known prehistorically and mentioned by Homer in 750 BCE, they can be seen easily with just your eyes. Most people, however, know them by a different name, the Seven Sisters. Even though they may not be aware of it, most people also know the Japanese name for the group, Subaru. The starry emblem found on the back of the car represents the star cluster.

Calling the Pleiades some variation of seven sisters is remarkably widespread. Indigenous people in Western Australia have a myth of seven sisters, Native American folklore refers to the cluster as seven children, and the ancient Greeks related the stars to the seven daughters of Atlas and Pleione, which is the origin of their group name.

The curious thing is that only six stars within the cluster can be seen easily with your eyes. Under suitable conditions, this number increases to nine, although theoretically there are 21 stars in the group visible to the naked eye. How many you see depends on your eyesight, atmospheric conditions and, unfortunately, light pollution levels. So, if it is possible to see either six or nine stars easily, how is it that the story of seven sisters is so prevalent across different cultures? It turns out the group represents seven sisters plus mum and dad, making nine easily visible stars. The two stars close together and off to the side of the main group are dad and mum, or Atlas and Pleione, as they are known in the Greek myth.

The cluster itself is a group of about 1,000 very young stars, all made from the same cloud of gas and dust about 100 million years ago and lying about 440 light-years away. The future of the cluster is estimated to be only another 250 million years, by which time they will have spread apart and no longer form a tight group.

Interestingly, even the faintest stars visible are still 40 times brighter than our own Sun would appear at a similar distance, and the brightest Pleiad, known as Alcyone, is about 2,000 times more luminous.

Exploring for yourself: How bad is the light pollution where you live?

Anyone who has looked at the sky from the middle of a city and then from a dark spot in the country will know that light pollution is a blight on modern civilisation. Determining just how bad light pollution is from where you live can be easily worked out by using a known group of stars and counting how many you can see. By knowing how many are possible to see from a dark site and counting how many stars you can actually see will indicate how much light pollution is interfering with your view.

Firstly, pick a cloudless night when the Moon is not visible, as moonlight will affect your answer. Also, wait for at least an hour and a half after sunset to make sure there is no residual sunlight.

Since the Pleiades is a group of stars most people know how to find, it is an ideal group to use. Using the following chart of the Pleiades, look at the group and cross off any stars you can reliably see. Repeat the exercise on several nights to help remove any interference from clouds, haze, and weather conditions. On the second chart, beside each star, is a number giving the star's magnitude, an astronomical brightness scale where the smaller the number, the brighter the star. Most people can see the six brightest stars in the group without too many problems, but under dark conditions and with good eyesight, at least three more should be visible. There are 16 stars shown in the chart, but to see all 16 is unlikely.

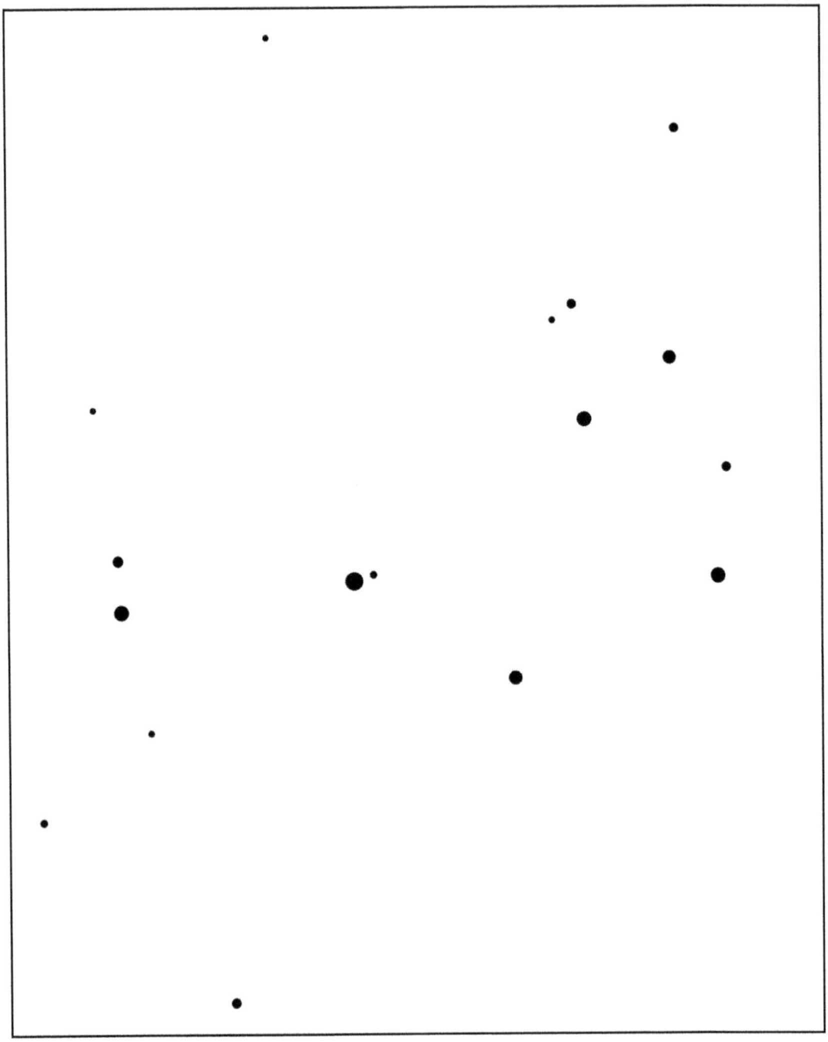

Star map of the Pleiades showing stars down to magnitude +6.5

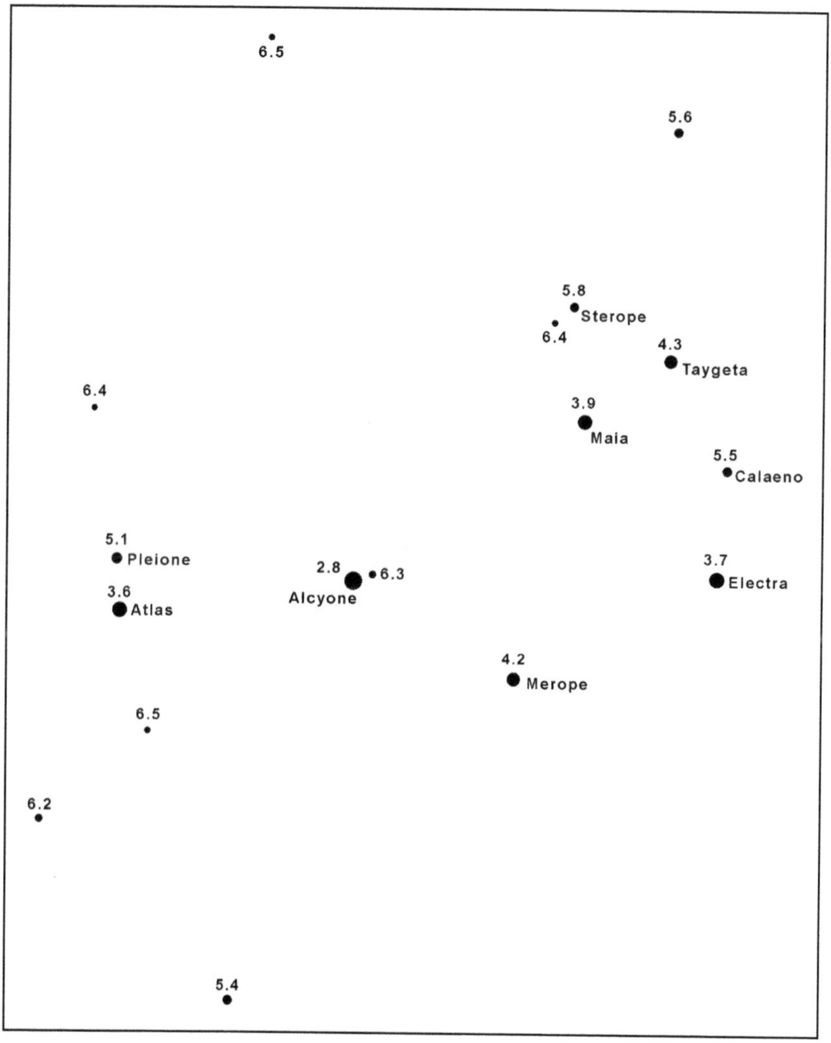

Star map of the Pleiades showing stars with magnitudes down to +6.5

After you have been through the process with your eyes, try it again with a pair of binoculars and the following chart of the Pleiades. This chart has the stars down to magnitude +10, the theoretical limit for a pair of standard 7x50 binoculars. For reference, the theoretical limit for your eyes is between magnitude +6 and +6½.

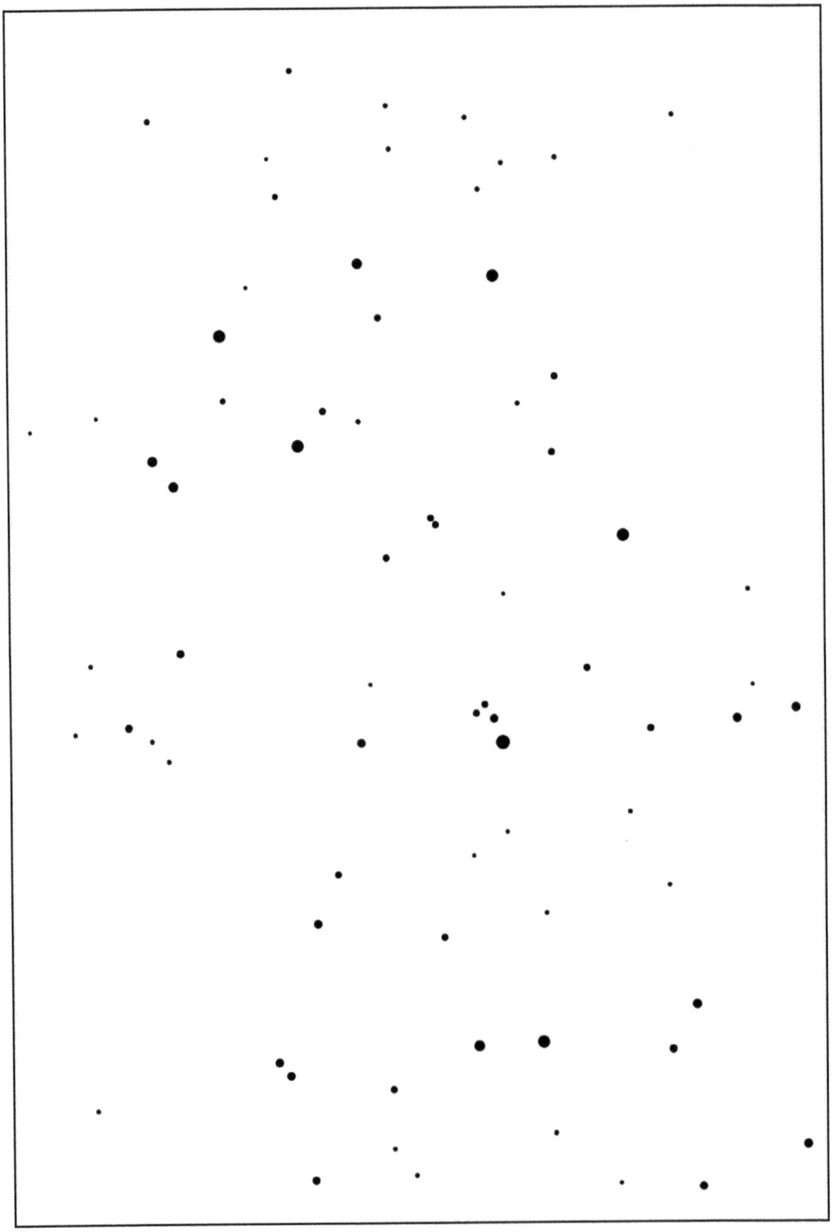

Star map of the Pleiades showing stars down to magnitude +10.0
(map is rotated 90 degrees counterclockwise to the maps shown on pages 176 and 177)

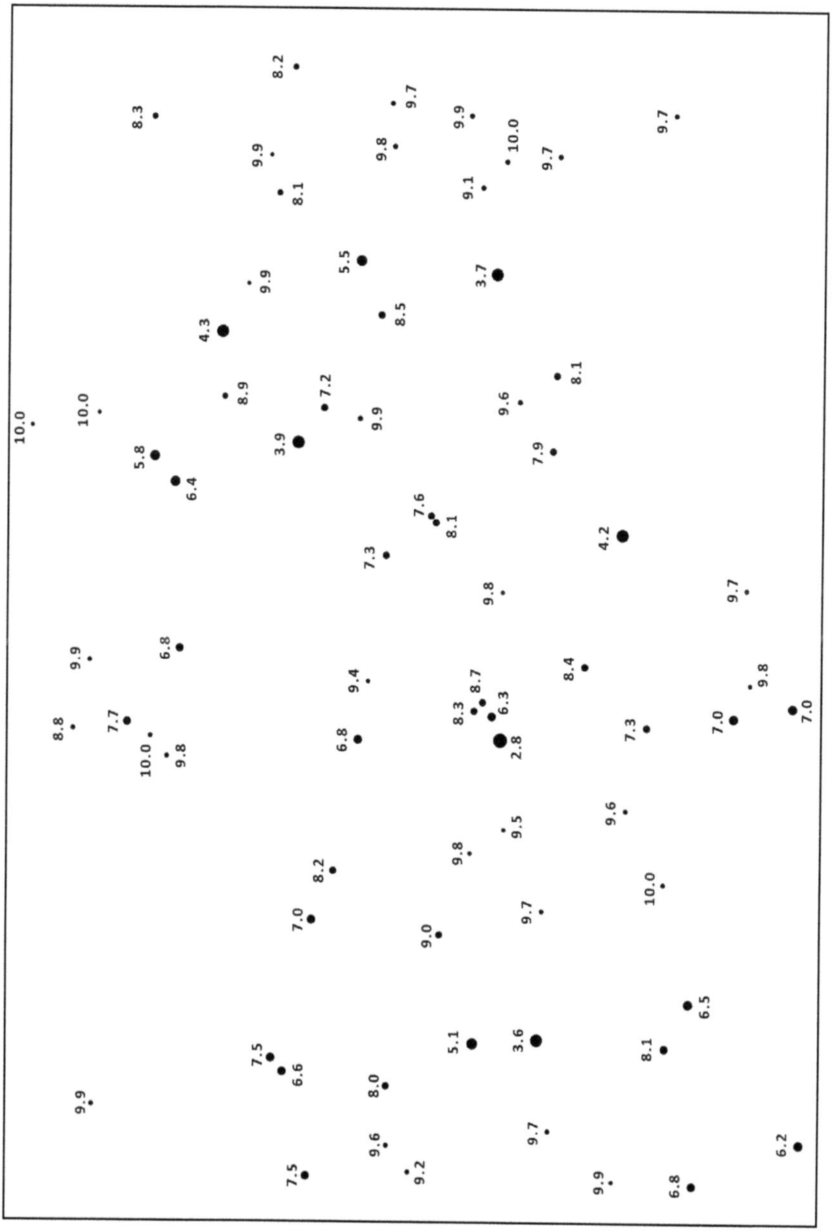

Star map of the Pleiades showing stars with magnitudes down to +10.0
(map is rotated 90 degrees counterclockwise to the maps shown on pages 176 and 177)

Since the Pleiades are only visible in the evening during summer, another well-known group of stars that can be used is the Southern Cross. This constellation is visible for most of the year, but only if you live south of the equator. However, being a larger area of the sky than the Pleiades means we can only use it for determining light pollution levels with our eyes. Once binoculars are used there are just too many stars visible to make it a useful group. So, if you want to use the Southern Cross as your main source or as a check to your Pleiades observation, the maps showing the stars theoretically visible to the unaided eye are also shown.

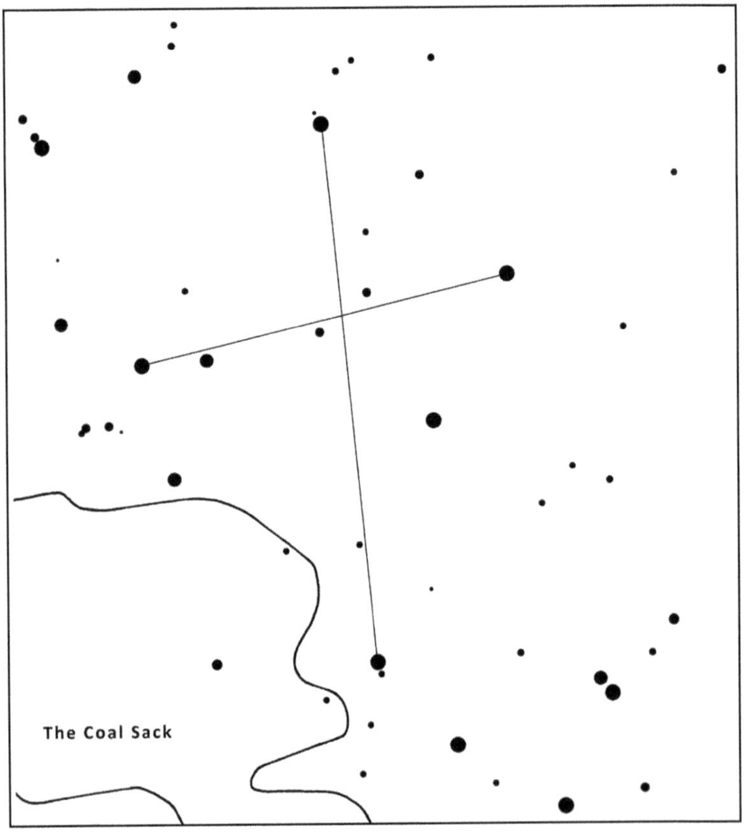

Star map of the Southern Cross showing stars down to magnitude +6.5.

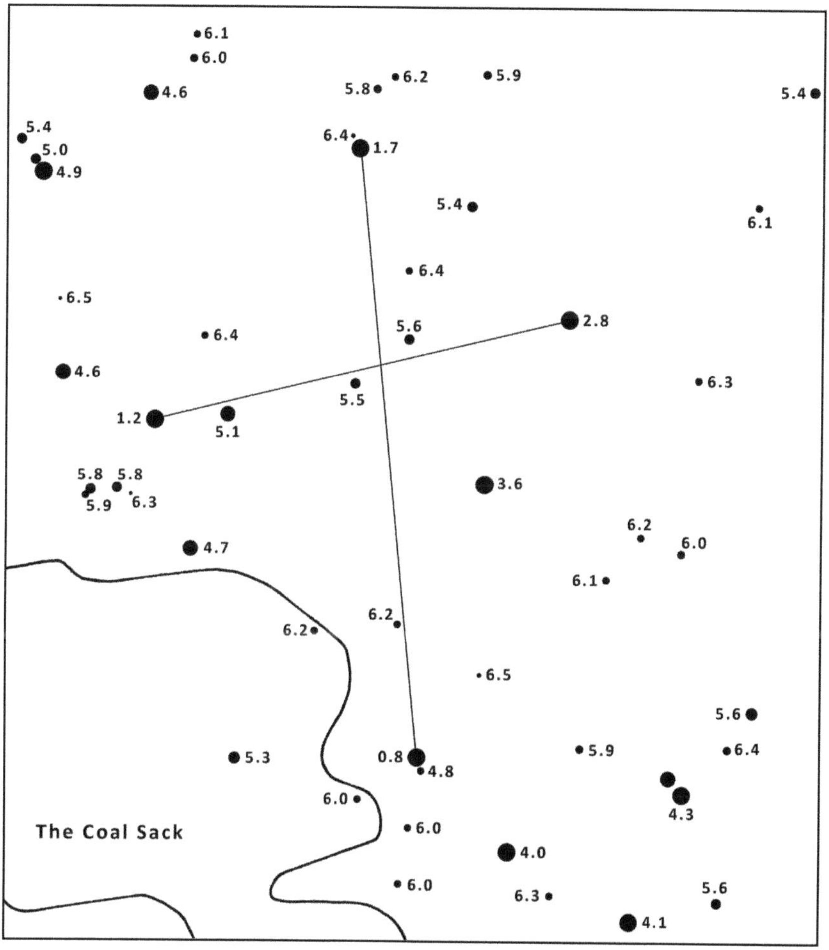

Star map of the Southern Cross showing stars with magnitudes down to +6.5.

Once you have made the observations, the brightest star on the chart you haven't been able to see will determine the limiting magnitude of stars visible from your location. With the following table, it is then possible to work out how many stars you can theoretically see in the sky and hence how many you are losing due to light pollution.

Magnitude Range		Number in Range	Cumulative Number
From	To		
-1.5	-1	1	1
-1	-0.5	1	2
-0.5	0	2	4
0	0.5	6	10
0.5	1	5	15
1	1.5	8	23
1.5	2	27	50
2	2.5	43	93
2.5	3	81	174
3	3.5	113	287
3.5	4	231	518
4	4.5	386	904
4.5	5	726	1,630
5	5.5	1,257	2,887
5.5	6	2,193	5,080
6	6.5	3,394	8,474
6.5	7	6,676	15,150
7	8	33,850	49,000
8	9	90,000	139,000
9	10	244,000	383,000
10	11	665,000	1,048,000

Table 2: Star counts

Sirius

Sirius is one of the jewels of the heavens. Lying close to Orion the hunter, but on the other side to Taurus, it is part of the constellation Canis Major, the big dog. Supposedly Orion's faithful companion who helps him fetch the game he is hunting, the constellation lends its name to an alternative title for Sirius, as it is sometimes known as the Dog Star.

As the brightest star in the night sky, Sirius stands out from the surrounding region. Although it is 21 times brighter than the Sun, its brilliance in our skies is primarily due to its closeness. At a mere 8.6 light-years away, it is the 5th closest star to us. Sirius is also gradually moving closer to the solar system and will increase slightly in brightness over the next 60,000 years.

Sirius is a double star with a difficult to see companion. The existence of this companion, called the Pup (what else would you call a companion to the dog star), had been inferred for a long time, but it wasn't discovered until 1862 when an American telescope maker found it while testing a telescope. This companion to Sirius was the first white dwarf ever discovered and is one of the few visible with a small telescope. White dwarves are the remnants of stars that have reached the end of their lives, shrunk to about the size of the Earth and became very hot and 200,000 times as dense as the Earth.

Sirius is also the only star known with certainty from the Egyptian records. Its hieroglyph, a dog, first appeared on the walls of temples about 3285 BCE. When Sirius rises at dawn, just before the Sun rises, it would presage the annual flooding of the Nile River. A significant event that deposited a rich layer of silt and made the land fertile again. Supposedly at this time, the heat of the Sun and that of Sirius (whose name means 'searing' or 'scorching') combined to produce sweltering weather. This is why the hottest part of the northern hemisphere summer became known as the Dog Days. This explanation of why it is so hot at this time isn't even remotely correct, but the term can still be found in use today. Also, it doesn't make sense if you live in the southern hemisphere as we have winter at this time.

Incidentally, it was the star Sirius that provided the name for the flagship of the First Fleet to Australia.

Spica

Spica is the 15th brightest star in the night sky and is not easy to locate without some aid, but it is one of the more fascinating stars. The name Spica means 'ear of wheat', as it supposedly represents a piece of wheat held by Virgo.

At a distance of 262 light-years, Spica is one of the closest blue giant stars to Earth. With more than 10 times the mass of the Sun, seven times the Sun's radius and about 12,100 times the brightness of the Sun, it truly is worthy of the title of Blue Giant and is one of the nearest stars to us

The constellation Canis Major (the big dog) and the brightest star, Sirius.

The constellation Virgo (the virgin) and its brightest star, Spica.

that has enough mass to end its life in a supernova explosion.

Spica is believed to be the star that enabled the Greek astronomer Hipparchus to discover one of the most important pieces of astronomical data about the Earth, the precession of the equinoxes. A temple at Thebes in Egypt was oriented towards Spica when it was built in 3200 BCE, but over time the precession of the equinoxes, caused by the slow wobbling of the Earth's axis, resulted in a noticeable change in the location of Spica relative to the temple.

Spica is a very close binary star whose individual stars orbit each other every four days. They are so close that resolving them through a telescope is impossible and they mutually distort each other with their gravity. As if that wasn't enough, the surface of the more massive star regularly pulsates outwards and collapses inwards over a matter of only four hours.

Located close to the path across the sky that the Sun, Moon and planets follow, Spica is sometimes occulted by the Moon and planets. The last planetary occultation occurred when Venus passed in front of the star back in 1783 CE. The next time will be in 2197 CE when Venus once again passes in front of it.

Alpha Centauri

Closer to home, trailing behind the Southern Cross each night are two very bright stars that show the way to the famous constellation. Of the two, the star furthest from the Cross is the one more worthy of our attention and a star we should get to know a little better. Lying at a distance of just over four light-years (approximately 41,250,000,000,000 kilometres) it is known as Alpha Centauri, the third brightest star in the night sky and our nearest known neighbour in space beyond the solar system.

Alpha Centauri is a multiple star system composed of three gravitationally bound stars. The two main stars are called Alpha Centauri A and Alpha Centauri B (not the most original names I know) and the tiniest star in the system is a red dwarf known as Alpha Centauri C, or Proxima Centauri.

The two brightest stars take about 80 years to orbit each other at an average distance of approximately 3,600,000,000 kilometres, a bit more than the distance from the Sun to the planet Uranus. Their closest approach to each other is about the same distance as from the Sun to just beyond Saturn, while their maximum separation is from the Sun to just beyond Neptune. Viewed from the Earth, their closest approach to each other occurred in 2016 CE, and they are now moving apart again.

Visible through telescopes as a magnificent double star, these two stars are remarkably sun-like, with Alpha Centauri A being a near twin of the Sun. At a calculated 1.227 times the size of the Sun, Alpha Centauri A is 10 per cent more massive and 50 per cent brighter. It is also known to have an eight year cycle of activity (the Sun's cycle is 11 years). Alpha Centauri B is 0.865 times the Sun's diameter, 10 per cent less massive and 50 per cent fainter than the Sun.

At 4.22 light-years away, Proxima Centauri is nearer to the Earth than the other two stars in the system, by a distance roughly 13,000 times the distance from the Earth to the Sun. This makes it the closest individual star to us, apart from the Sun. Proxima orbits the other two stars in a huge circle that takes between 500,000 and 1,000,000 years to complete. The uncertainty is due to it having moved only a tiny amount in its orbit since it was discovered, which makes determining its exact orbital period extremely difficult.

Discovered in 1915, Proxima Centauri is one of the least luminous stars ever found and is 13,000 times fainter than the Sun. If it replaced the Sun, it would be only 45 times brighter than the full moon and would appear to be only 1/20 its diameter. Proxima is also known to have a planet that is 1.3 times the Earth's mass, orbits Proxima at 1/10 the distance of Mercury, has an 11 day year, and lies within Proxima's temperate zone. In the same way the Moon only shows one side to the Earth, Proxima's planet is more than likely tidally locked, only ever showing one side to Proxima Centauri.

With one of the most substantial motions across the sky, in about the year 6000 CE, the Alpha Centauri system will have moved enough to bring it close to the other pointer star, Beta Centauri. They will then become an incredible visual binary star.

The constellation Crux (the Southern Cross) and the two pointer stars, Alpha and Beta Centauri.

The constellation Crux (the southern cross) on the left,
and the bright nebula, Eta Carina on the right.

Eta Carina

The last star I want to mention is also one of the most fascinating. Prominent in the southern sky is one of the richest constellations, Carina, the keel. Originally part of the much larger constellation Argo Navis, the ship, Carina lies along the Milky Way and contains some of the best objects to look at. In particular, it has two of the most interesting stars, Canopus (the second brightest star in the night sky) and Eta Carina.

Eta Carina is a star located within the star-forming region known as the Eta Carina Nebula. Probably formed less than a million years ago, it has at least 100 suns worth of material, making it possibly the most massive star in the Milky Way galaxy. It pumps out over 4,000,000 times the energy of the Sun and will end its life in a hypernova explosion. Being such a massive star, Eta Carina uses its available fuel extremely quickly and consequently has a short life span, probably no more than 1,000,000 years. That means it is already approaching the end of its life.

First catalogued by Edmund Halley in 1677 CE, Eta Carina increased in brightness by six times before it faded back to its original brightness by 1811. It then steadily increased in brightness again until, in 1843, it reached a brilliance nearly 100 times its original. At that time, despite its distance of 7,500 light-years, it outshone all the stars in the sky except for Sirius.

Since 1843 Eta Carina has faded until it is now only just visible to the unaided eye. This fading act is thought to be due to the material it ejected in 1843 blocking our view of the star in the middle. In 1999 it once again doubled in brightness.

The outburst of 1843 could be a precursor to the star exploding soon. Exactly when it might happen is open for debate. Some astronomers think it won't explode for about 100,0000 years, while others believe it could happen at any moment. When it does explode, it will be easily visible during the daytime and bright enough to read by at night. The effects of the explosion may be felt by the Earth, although the atmosphere should protect us from the worst consequences. It might, however, be a different story for any astronauts and satellites. My apologies to the astronauts, but I hope it happens sometime soon as I would really like to see Eta Carina explode.

The Eta Carina nebula.

CHAPTER 17

Constellations

What are constellations?

Constellations are arbitrary groupings of unrelated stars that have been lumped together because someone once thought they seemed to make a picture. Of course, this is a considerable oversimplification, as there are numerous reasons why humans wanted to make pictures among the stars.

Thousands of years ago, when looking into a dark night sky, people were mystified by the twinkling panorama above their heads. To make sense of this, and impose some order on the apparent chaos of the sky, they invented patterns out of the stars. These patterns embodied mythological characters, animals, gods, heroes and monsters, all playing out their roles in the night sky. Since they were a projection of people, deities and creatures, these celestial patterns were then used as a storybook to teach morals and traditions to younger generations, long before the invention of books or the internet. Plus, once there were recognisable star patterns imposed on the heavens, they could then be used for more practical purposes, such as navigation, a calendar, or a clock. But perhaps most importantly, in the beginning, they made the fearful night sky a little bit less scary.

Every civilisation around the world made patterns in the sky, each with their own particular pictures and stories. But around 100 years ago, the International Astronomical Union (the worldwide governing body of all things astronomical) decided there should only be 88 official constellations, all with well-defined boundaries so that no part of the sky was outside of one. Kind of like a vast three-dimensional 88 piece jigsaw puzzle surrounding the Earth. These official patterns are based primarily on a list made in 150 CE by the Greek astronomer Ptolemy, although he didn't invent these constellations. Most date back to the Babylonians and Sumerians about 4,000 years ago.

As constellations are the invention of human imagination and not nature, Ptolemy's original list of 48 constellations was extended in a burst of constellation creation between 1500 CE and the mid-1700s to yield the

The southern Milky Way from Alpha Centauri to Eta Carina.

now recognised 88 patterns. Some of the more recent (mainly southern hemisphere) constellations were created to celebrate then recent inventions. A few were attempts by their inventors to immortalise themselves for all eternity. A breakdown of the 88 constellations shows there are 14 humans (or part thereof), five fish, nine birds, 23 other animals, 29 inanimate objects, and eight mythological creatures represented.

In case you were wondering, the largest constellation is Hydra, the many-headed snake fought by Hercules. The smallest is Crux, the Southern Cross. The constellation with the most number of bright stars is Orion. The most number of visible stars is Centaurus, and the most visible stars per square degree is Crux. Make sure you remember these facts as they will surely come in handy at a trivia night at least once.

The constellation Capricorn (the sea-goat).

The constellation Grus (the crane).

Why do we keep the constellations?

One of the biggest disappointments when looking for constellations is that they bear little to no resemblance to the figures they are meant to represent, except for Scorpius. Looking like a giant fish hook of stars, it's not too hard to create the body and tail of a giant scorpion. Another is Crux, the Southern Cross, and just nearby, Triangulum Australe, as it's hard to go wrong when you call three stars in the south the Southern Triangle.

So, if they don't look like the pictures they are supposed to represent, and we no longer believe in centaurs, sea monsters et al, why do we still keep the constellations today? I can think of a couple of good reasons.

Firstly, constellations are handy to find your position in the sky in the same way that countries are helpful to know where you are on Earth. If you say you were looking at something in the constellation of Centaurus then immediately someone else knows approximately the area of sky to start searching. In other words, they are a useful means of dividing the sky into conveniently sized chunks.

The second reason is simply that they are fun to have.

Why is it we can join the stars and make pictures?

A bigger question than why we keep the constellations is why we can make constellations in the first place. This might seem like a trivial question, but it highlights a crucial aspect of the universe, it's big. If the universe is so big and the stars so far away from us, how is it we can make stick figures out of them? It may seem obvious that we can join the dots when we look at them, but if we consider that the stars are varying distances from us, it isn't quite so clear cut. The question should then become why we don't notice one star is further away than another?

To tell distances of more familiar objects, such as a tree, we rely on having two eyes. Because our eyes are a small distance apart, they see things from a slightly different angle. Put your finger in front of your eyes and alternately open and close one eye at a time, and you will notice this quite easily. The finger appears to change position against the background as each eye looks at it from a slightly different angle. The brain interprets this difference and gives us a sense of distance to the object. The less difference between the views of each eye, the further away it must be.

When it comes to stars, they are so far away that our eyes are not able to notice a difference in position from one eye to the other. This means we cannot tell how far away they are, or which star is closer than another, using just our eyes. To notice any difference at all (and it would still be minuscule) you would need to have one eye on one side of the Earth's orbit and the other eye on the other side.

Since no one has eyes that distance apart it is not possible to tell how far away the stars are, which means when we look into the night sky the stars all appear to be the same distance from us. This gives the impression of the stars being on the inside of an enormous sphere surrounding the Earth and allows us to draw pictures to our heart's content. It is also why we can create a coordinate system for the sky.

The human mind has an uncanny knack for making order out of apparent randomness (in case you were wondering it is called pareidolia) and it is only natural that we make pictures out of these seemingly random dots in the sky. And although the stars themselves have no connection beyond the imaginary lines the mind has created, the patterns produced make the night sky a little bit more interesting.

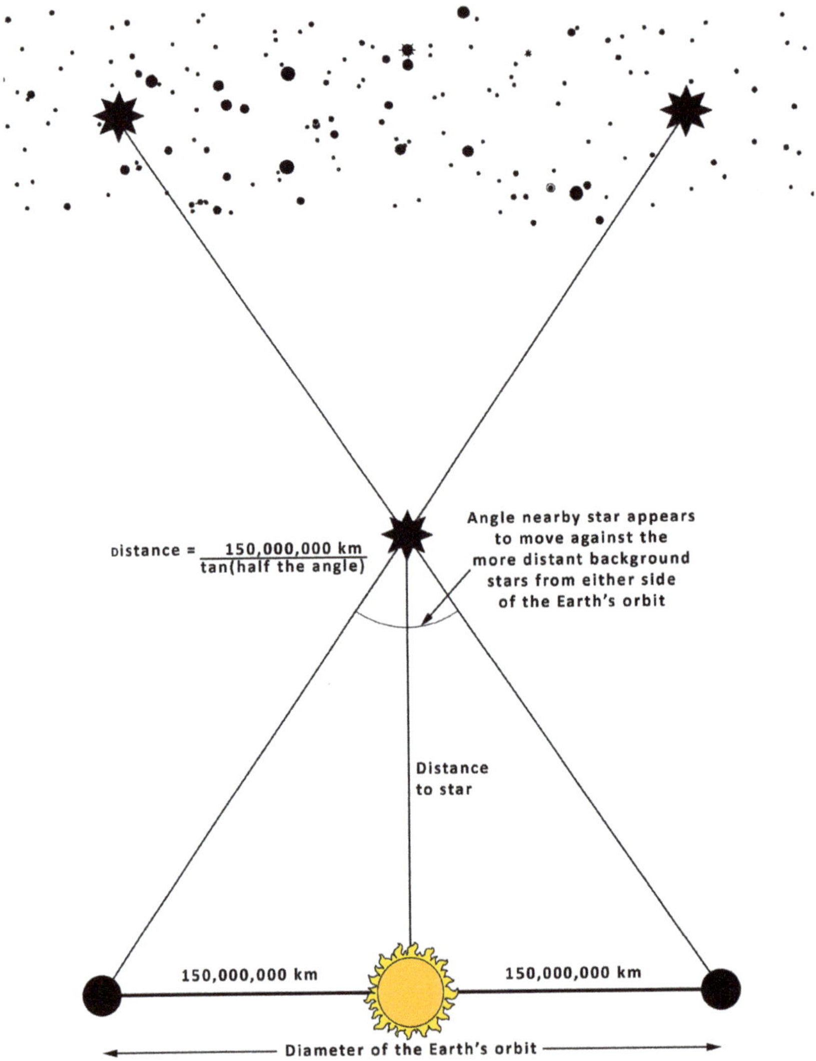

Determining the distance to nearby stars using parallax.

However, the stars are, of course, in three dimensions, with all of them moving in different directions. Throughout a handful of human lifetimes we won't notice any difference, but in a few thousand years the current patterns we see will be destroyed. Our distant ancestors will not see the same pictures we currently see, but undoubtedly by then the constellations will have morphed into figures that make sense to them.

Exploring for yourself: The Southern Cross in 3D

When we look at the Southern Cross, the five stars appear to form a cross, or kite shape, and although the stars look close together in the sky, they are hundreds of light-years apart. The shape they form is only how they look when seen from Earth. If we could view the stars that make up the Southern Cross from somewhere else in space, far from our solar system, they may not look anything like the familiar kite shape. This can be easily demonstrated by making a three dimensional model of the Southern Cross, allowing us to view the stars from all directions. The table below gives the distances to each of the five main stars in the Southern Cross, plus the two Pointer stars, and the following star map shows the names of each star in the Southern Cross for reference.

To make the model, you need a Styrofoam, or thick cardboard, sheet stiff enough and large enough to draw a reasonable sized Southern Cross onto it. You will also need a needle and thread and some foam balls to act as stars.

Star	Distance (in light years)	Relative distance (alpha Centauri = 1cm)
Alpha Centauri	4.2	1
Beta Centauri	525	125
Alpha Crucis	321	76
Beta Crucis	353	84
Gamma Crucis	88	21
Delta Crucis	364	87
Epsilon Crucis	228	54

Table 3: Crux star distances

Template of the Southern Cross and Pointers.

Using the provided star map of the Southern Cross as a template, draw the stars onto the sheet. Thread the needle with enough thread for some adjustment and stick it through the foam ball. Then stick it through one of the stars marked on your template. Pull until the thread is as long as the suggested length given in Table 3 for that particular star. Using tape to secure the thread to the back of the sheet, repeat for the remaining stars. My suggestion is to do this while the template is suspended. That way the foam balls will hang straight down due to gravity and you won't get the threads in a massive tangle like I did the first time I did it.

Once the three-dimensional model is completed, look at it from underneath and you should see the familiar Southern Cross shape. Look from the side, however, and it will appear completely different.

Exploring for yourself: The Saucepan in 3D

Not everyone on Earth can see the Southern Cross, but we can do the same exercise for a group of stars that everyone is familiar with, the Saucepan, or Orion's Belt and Sword. The star distances and star map for the Saucepan are given below. Follow the same instructions as for the Southern Cross to create a three dimensional model of the Orion's Belt area.

Star	Distance (in light years)	Relative distance (45 Orionis = 10cm)
Eta Orionis	901	24
31 Orionis	456	12
Delta Orionis (Mintaka)	916	25
Epsilon Orionis (Alnilam)	1342	36
Zeta Orionis (Alnitak)	817	22
Sigma Orionis	1148	31
45 Orionis	371	10
42 Orionis	786	21
M42 (Orion Nebula)	1400	38
Iota Orionis	1326	36

Table 4: Saucepan star distances

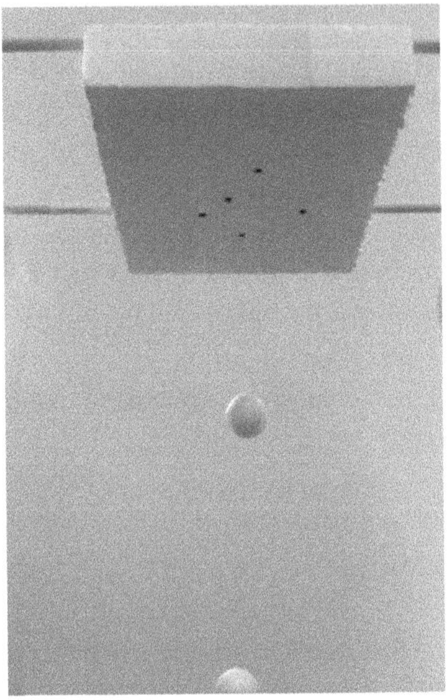

Making a model of Crux.

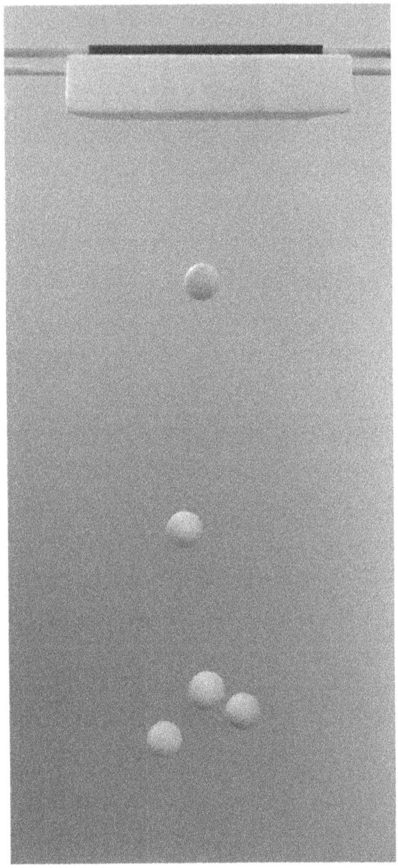

Looking at the Crux model from the side.

Crux model from the top.

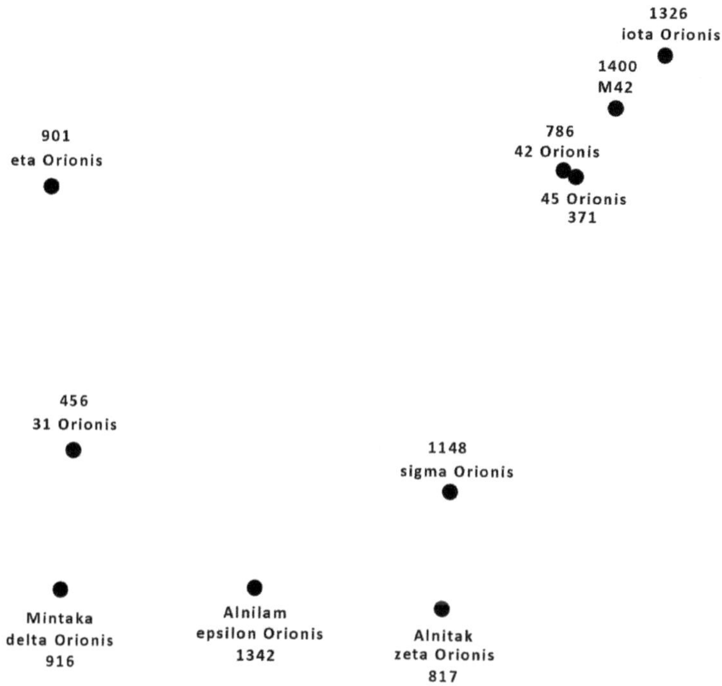

Template of The Saucepan.

Note: the middle point of light in the handle of the Saucepan does have stars in the area, but the primary source of light comes from the Great Nebula in Orion. Consequently, I have given the distance to it, rather than the stars.

Zodiac

The word Zodiac means 'Circle of Animals', but a quick check of constellation names shows that one doesn't fit the description. However, it all makes sense when you realise that the stars of Libra, the balance scales (and the odd one out), were originally part of Scorpius. Created after the constellations were collectively called The Zodiac, Libra was made by snapping off the claws of the scorpion and turning them into a separate constellation during the reign of Julius Ceasar. They did this because at the time the Sun appeared within this constellation on the equinox and the balance scales were supposed to symbolise the equal lengths of the day and night at this time of year. Although Libra marked the position of the equinox 2,000 years ago, the Sun now resides within Virgo at this time. Incidentally, the two brightest stars in Libra have the best names ever, Zubenelgenubi and Zubeneschamali. Because they were initially the claws of the Greek scorpion, the names originally come from the Arabic for southern and northern claw.

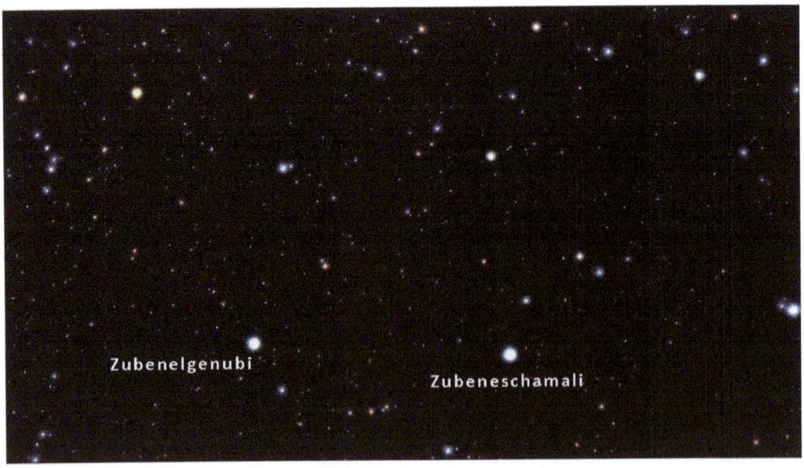

The constellation Libra (the balance scales).

It is a common misconception that the Sun, Moon and planets are only ever found in the Zodiac constellations and that this is what sets them apart. This isn't strictly true.

When a few planets are in the sky at the same time, you will notice that they all seem to lie in a straight line. This imaginary line is called the ecliptic and it may seem strange to have the planets in a line until we realise that all the planets move around the Sun in roughly the same plane. Since we live on one of the planets, when we see the others, we're looking side on to this plane and consequently only ever see them in a straight line across the sky. Picture the planets rolling around on the surface of a table with the Sun at the centre. It also explains why the Sun and Moon follow the ecliptic as well. However, the Moon and planets are not positioned exactly on the ecliptic because they're not precisely in the same orbital plane as the Earth. They tend to lie within several degrees of it and form a narrow strip across the entire sky, which we call the Zodiac. The ecliptic runs along the exact middle of the Zodiac, and the band extends for nine degrees either side of it.

If we look at which constellations the ecliptic passes through, we end up with the traditional 12 zodiac constellations, plus one. These constellations are the only ones through which the Sun passes. Remarkably, the Sun spends more time in the 13th constellation, Ophiuchus (where it resides for 18 days between 30 November and 17 December), than in the more commonly known Scorpius (where it only spends seven days between 23 November and 29 November). Yet somehow Ophiuchus is not considered a member of the Zodiac and must defer its position to Scorpius. These are the only constellations the ecliptic passes through, but if we extend the list to incorporate all of the constellations the zodiac band intercepts, the number is increased by another eleven.

Since the Moon and planets don't travel around the Sun in the same plane as the Earth, they are often positioned either just above or below the ecliptic. This means that they appear within the boundaries of some of the non-zodiacal star patterns. Add in the Earth's axis wobbling over a 26,000 year period, and it makes a total of 24 constellations that can contain either the Sun, the Moon or a planet, a reasonable percentage of the 88 constellations up there.

Crux and finding your direction

Easily visible from anywhere in the southern hemisphere is perhaps the most famous constellation of all, the Southern Cross, or Crux as it is officially known. Although it is the smallest of all the 88 constellations, it is made up of some of the brightest stars in the heavens. Acrux, the brightest one, is the 14th brightest star in the entire sky.

Its size, and the fact that it looks more like a kite than a cross, can make it difficult to find for the first time. Fortunately, there is an easy way to locate the Southern Cross and be confident you have it. Each night, as it passes across the sky, the Southern Cross has trailing behind it two bright stars known as The Pointers, so named since they seem to point to the top of the Cross. The line between them does, in fact, just miss the top of the Cross, but it's close enough. These two pointer stars, Alpha Centauri and Beta Centauri, are very bright with both in the top ten list, making them virtually impossible to miss.

Historically the Southern Cross has had an interesting life. The stars in the Cross were known to the ancient Greeks but they regarded them as the hind legs of the constellation Centaurus, the centaur, which today surrounds the Cross. It wasn't until the year 1516 CE that it was first described as a cross by the explorer Amerigo Vespucci (whom America is named after) and not until later that century that it was adopted as a separate constellation by astronomers.

Long before compasses, it was possible to navigate by using the stars, so, in case you get lost in the desert or out at sea without a compass, how do you do this?

If you keep track of the stars throughout the night you will notice that they all seem to rotate about a fixed point in the sky. If you watch long enough, you would see they do one complete circle in one day. This motion around a single point is due to the Earth spinning and consequently the point about which all the stars circle is, therefore, directly above the Earth's rotation axis, the south pole. If we can locate this point in the sky, known as the South Celestial Pole, it is then possible to work out which direction is true south by dropping an imaginary line from this point straight down to the horizon. Once you know which way is south, the other compass points are then easy to determine.

In the Northern Hemisphere, a relatively bright star called Polaris is close to the North Celestial Pole, so finding north in the Northern Hemisphere is reasonably straightforward. Unfortunately, here in the southern hemisphere, finding the South Celestial Pole isn't quite so easy, as we don't have a bright star close to this point. To get around this problem, we need to be a bit more inventive and use the Southern Cross.

There are two ways we can use the Cross to find the South Celestial Pole. But first, we need to think of the Cross as a kite. If we do, then like most kites, it has a tail at the bottom. If we think of the constellation in this way, it doesn't matter what orientation you observe the Southern Cross to be in, the bottom of the kite shape will be easy to work out.

The first way of finding the South Celestial Pole is to extend the long axis of the Southern Cross from the top of the kite through the bottom of the kite (regardless of orientation) by 4½ times its length. It isn't perfect, but it gets you close to the South Celestial Pole. Certainly close enough if you ever get lost. Alternatively, extend the long axis of the Cross from

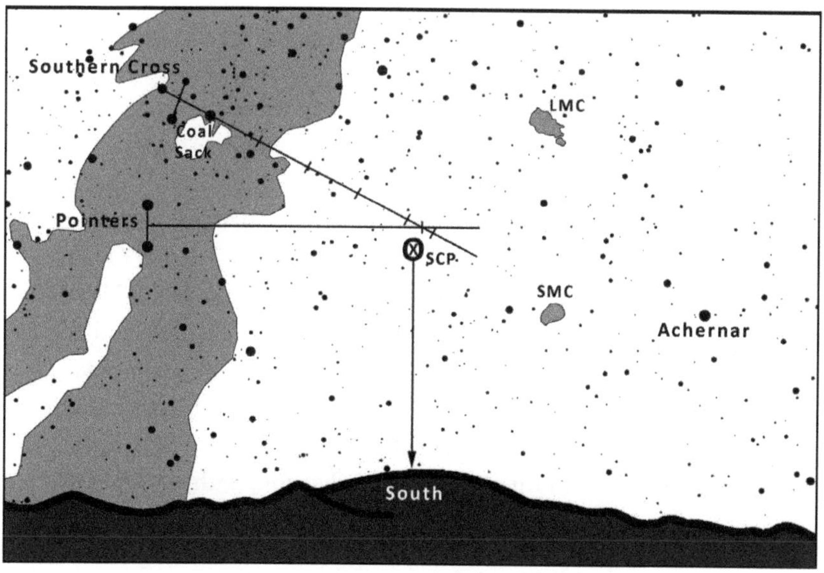

Methods for finding south using the Southern Cross.

the top of the kite through the bottom of the kite for as far as you like. Then, starting halfway between the two pointer stars, draw another line at 90 degrees to the line joining the two stars towards the first imaginary line. Where these two lines cross is roughly the South Celestial Pole. Again, it isn't perfect, but it's good enough to use. Once you have mastered these methods, there's no longer any excuse for not knowing your direction. All you have to do is wait until dark and you will then know precisely which way to go.

Exploring for yourself: Finding the South Celestial Pole

Roughly finding south is easy, use a compass. But this gives you magnetic south, not true south. Magnetic south is the direction of the Earth's magnetic South Pole and true south is the direction of the Earth's rotation pole and this is what we are referring to when we talk about the actual South Pole. The two are not in the same place and the magnetic pole continually changes position.

In the north, the star Polaris sits (for the moment) near the North Celestial Pole. In the south, however, it is more of a challenge as the closest naked eye star to the South Celestial Pole is Sigma Octantis, a faint star, barely on the edge of visibility, coming in around the 4,000th brightest mark.

If Sigma Octantis is virtually impossible to see, how can you find the exact South Celestial Pole? As a starting point, the approximate methods mentioned earlier using the Southern Cross are all you need to find the rough location of the South Celestial Pole. Once you have the general location, use binoculars and the following star map to identify Sigma Octantis. Look for the patterns the stars make around Sigma Octantis to orientate between the map and what can be seen through the binoculars. Sigma Octantis isn't precisely at the South Celestial Pole, but find the star and it is then possible to locate the blank spot that is the exact pole.

Because sigma Octantis is so faint, the first time you try to find it, do so from a dark site, away from any light pollution. Once you get the hang of it, you can start venturing closer to civilisation, with its light polluted skies, and still be able to locate it.

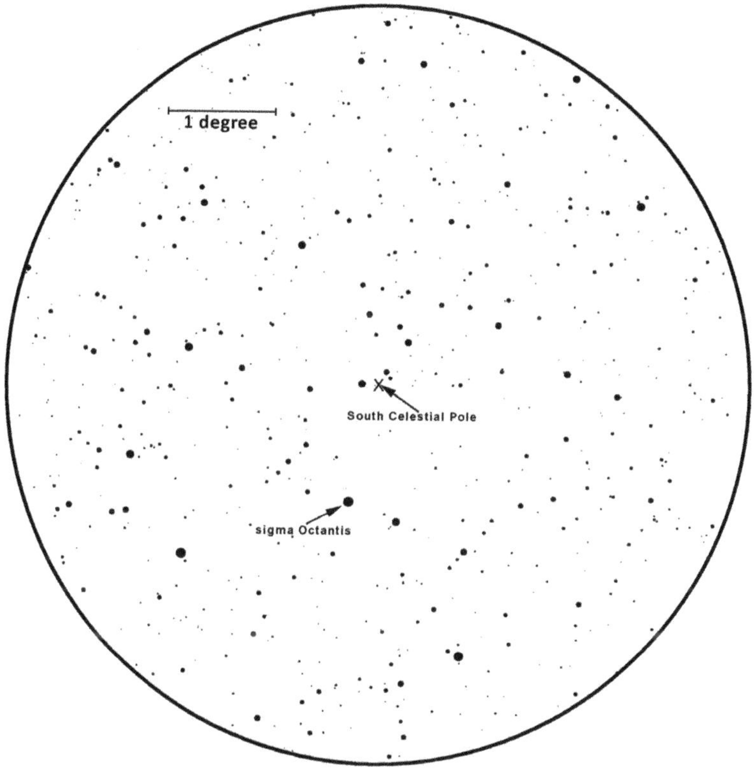

Locating the South Celestial Pole.

Exploring for yourself: Tracking a star and the Southern Cross

Since the South Pole is at the Earth's rotational axis, if you stand at the exact pole you will remain stationary while everything rotates around you. The closer something is to the pole, the smaller the circle it makes around you, the further away it is, the larger the circle. Since all circles are completed in one day, anything closer to the pole will move slower than something further away, simply because they have less distance to travel in the same amount of time. The same is true of the stars. All stars appear to rotate about the point directly above the South Pole due to the Earth's rotation.

On any clear night, go outside and make a note of exactly where you are standing and what direction you are facing. It is not necessary to know the actual compass direction, only the direction you are looking relative to a prominent landmark. Locate any bright star and note it's position in relation to your landmark. About an hour later return to the same place and you will notice the star has moved. The closer the star is to the South Celestial Pole, the less it will have moved and the further away from it, the more it will have moved across the sky.

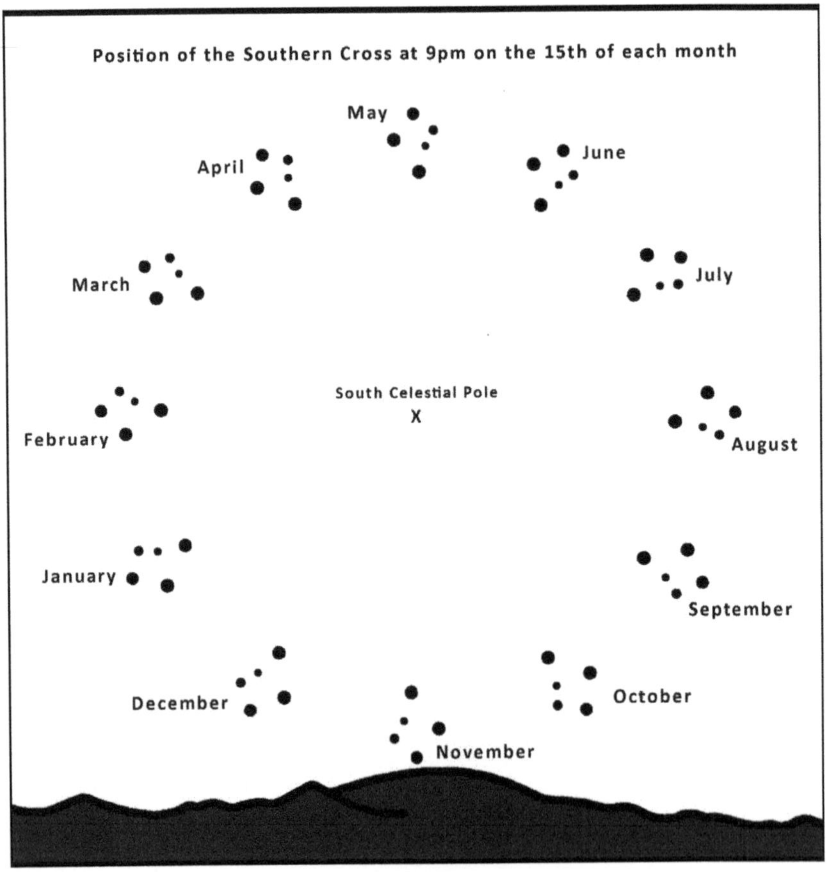

The seasonal (and nightly) rotation of Crux.

To specifically observe the motion of stars around the South Celestial Pole, locate the Southern Cross and Pointers and watch them for a few hours. Due to its distinct shape and orientation, it easily demonstrates that all stars rotate around the South Celestial Pole. After not quite 24 hours, they will have completed one full circle.

The rotation period isn't quite 24 hours, so if you observe the Southern Cross each night at the same time according to your watch, it will have moved slightly around the South Celestial Pole. After one year it will be back in the same position at the same time, but because of this slight change in position each night, where you find the Southern Cross in the sky depends on the time of year you look. People have often mentioned to me that the Southern Cross has disappeared when, in fact, it is just hanging upside down and low to the horizon. As the Cross changes orientation, it is also easier to refer to the star Alpha Centauri as the pointer furthest from the Cross. Depending on the time of night and time of year it can be the top, right, or bottom pointer, but it is always the pointer furthest from the Southern Cross.

Asterisms

Finding the official constellations in the sky is difficult. It would be so much easier if you could make your own. So the question is, are you able to make your own constellations, and the answer is a resounding yes. They won't be officially recognised, but if the pictures you create make it easier for you to find your way around the sky, then go for it.

The official constellation pictures were made so long ago that some are unfamiliar to our modern lives. After all, how many times have you come across a half-goat half-fish (Capricornus) in your travels? This unfamiliarity makes the constellations tricky to locate and recognise, so some people make their own pictures from the stars that are relevant to today and easier for them to find night after night. Since these personal pictures are not part of the official constellation list, they are known as asterisms.

Perhaps the best known asterism is The Saucepan (or The Pot), made out of the stars that make up the belt and sword of Orion, the mighty hunter. This is a typically Australian and New Zealand asterism, but since these stars are visible from anywhere on Earth, this particular group is known by a multitude of names around the world.

Another familiar asterism is the False Cross in the constellation Vela. So named because at times when the Southern Cross is low to the horizon, the False Cross is usually high in the sky and often gets mistaken for the real Cross.

My personal favourite asterism, however, is the shopping trolley, made from the stars of Corvus, the crow. You have to view it from a certain angle but give it a go and see if you can work it out. A shopping trolley is a lot easier to see in those stars than a crow.

Another famous asterism is known as The Teapot. Made from the stars that make up the bow and arrow of Sagittarius, it is visible almost directly overhead of an evening during the Southern Hemisphere winter. Looking for a centaur holding a bow and arrow is challenging, to say the least, but a teapot located just near the tail of Scorpius is relatively easy to find and is a good starting point to locate many remarkable nearby things in the Milky Way.

The constellation Corvus (the crow) looks more like a shopping trolley than a bird.

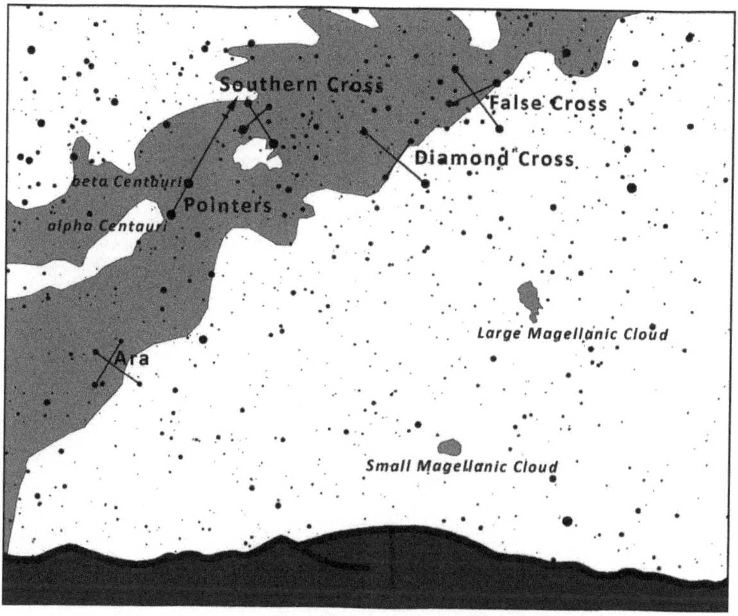

There are numerous false crosses in the sky not far from the real Southern Cross.

False crosses of the southern sky.

Sagittarius (the archer) makes a better teapot than a centaur with a bow and arrow.

The Saucepan.

The Saucepan in detail

The stars that make up the Saucepan are some of the brightest stars in the sky, and the asterism contains some fascinating objects.

Within the handle of The Saucepan (the sword of Orion) is one of the best, brightest and most easily seen nebulae, the Orion Nebula. Visible to the unaided eye, stars are currently forming within this vast cloud of gas and dust.

Between the handle and the base of the Saucepan is a faint star called Sigma Orionis that is made up of five members in an intricate gravitational dance around each other. Two of the stars orbit each other every 170 years while the other three are in unstable orbits and long before they die will be gravitationally flung out. They will end their lives as white dwarves wandering the galaxy.

Attached to one of the stars making up the base of the saucepan is the famous Horsehead Nebula. Unfortunately, it is incredibly faint and not easily seen outside of a photographic image.

Perhaps the most interesting object is the middle star of the base, known as Alnilam. It is the most massive star we can see with the unaided eye. Alnilam is not a binary star, so measuring its mass is difficult, but by comparing it with other stars, we estimate that it weighs an astonishing 80 thousand trillion trillion tonnes, some 40 times more than the Sun. It is also hundreds of thousands of times more luminous than the Sun but fortunately lies about 2,000 light-years away. Because it is more distant than almost every other star visible with our eyes, instead of being overwhelmed by Alnilam's brightness, we see it as the 29th brightest star in the sky.

CHAPTER 18

Further afield

The Milky Way

The Milky Way is one of the most spectacular objects in the night sky. Stretching from horizon to horizon, it passes directly overhead during Winter. Although other cultures had stories explaining this luminous band, it is the ancient Greeks, who saw it as milk, squirted across the sky by one of the gods, that gives us the name we commonly use.

The Milky Way, however, is not a cosmic case of spilt milk but the bulk of our galaxy, a collection of hundreds of billions of stars. Our star, the Sun, lies about two-thirds of the way out from the centre. Since our galaxy is a flat spiral shape (looking a bit like a fried egg) and we're looking from within it, we see the main part of the galaxy (the white part of the egg) as a band across the sky. This luminous band looks solid because the stars that make it up are so far away and so numerous that they blur together. If we look towards the central bulge of the galaxy (the yolk), we find the band of the Milky Way widens in the constellations Scorpius and Sagittarius.

Using nothing more than our eyes, we can see that throughout the Milky Way there are dark lanes and patches. These dark regions are not holes in the Milky Way, but dark clouds of gas and dust that block our view of the stars behind them. With a bit of imagination, out of these dark lanes it is possible to see an emu lying on its stomach with its head beside the Southern Cross and its body in Sagittarius.

The location of the galactic centre was initially determined by noting that all the visible globular star clusters seem to surround our galaxy. By working out the central point of this swarm, it was then possible to work out the centre of the galaxy. Astronomers discovered at this central point a tiny object known as Sagittarius A* (pronounced Sagittarius A-star), which, as far as we can tell sits at the exact centre of our galaxy. Everything else circles around it, but Sagittarius A* stays still. Astronomers have estimated both the mass and size of Sagittarius A* and the results are amazing. Sagittarius A* is millions of times more massive than the Sun, but so small that it would fit inside the orbit of Mercury. That means the only thing it could conceivably be is a supermassive black hole, an object whose density is so high and gravity so intense that anything silly enough to stray too close is sucked into it, forever. As the name suggests, not even light can escape from a black hole. So while the Milky Way's central black hole is itself presumably black, gas spiralling into it produces the vast amounts of light and energy we see as Sagittarius A*. So next time you are outside gazing at the Milky Way, keep in mind that you're looking at a barred spiral shape of billions of stars swirling around a supermassive black hole and wonder at what a magnificent universe we inhabit.

Incidentally, if you are in a really dark location and marvelling at the spectacular Milky Way visible, then not long after sunset or before sunrise you might notice what looks like a second Milky Way reaching up from the horizon. Our galaxy isn't splitting itself in half. Instead, you are looking at something known as the Zodiacal Light. The solar system is full of fine dust particles which scatter the Sun's light. This scattered light is very faint so it is easily drowned out by moonlight or light pollution, which is why it can only be seen from a dark location. Since the dust is in the same plane as the planets and the planets only pass through the Zodiac constellations, this scattered light can only be found along the zodiac as well, hence its name.

Visible Exoplanet Stars

Like most constellations, unless you have some help, it is difficult to see the famed twins of mythology from the seemingly random stars that make up Gemini. It is, however, straightforward to find its two brightest stars, Castor and Pollux. These two stars represent the heads of the boys and conveniently lie side-by-side in a faint part of the sky.

Castor, the more northerly of the two stars, is a remarkable system of six stars engaged in an intricate gravitational dance. With a moderate sized telescope you can see that the point of light visible with your eyes is, in fact, three stars. Each member of this trio, however, turns out to be a very close pair, too close to be split with any telescope.

The other star, Pollux, is one of the brightest stars in the sky confirmed to have at least one planet orbiting it, a planet a bit over two times the mass of Jupiter.

That raises the question, are there any other stars visible with the unaided eye that have planets orbiting them?

Two stars make up the chest of Leo, the lion. The brighter star, Regulus, is exciting in its own right, but the other star is called Gamma Leonis, or Algieba, and has at least one planet, maybe two, orbiting it. Close to Orion is the V shape of stars making up the head of Taurus, the bull. At the end of one of the arms of the V is the bright, reddish coloured star Aldebaran, home to one known planet with a mass about 6½ times that of Jupiter. That's a big planet. Amazingly, at the end of the other arm of the V shape is the star Epsilon Tauri, which has a planet with a mass 7½ times that of Jupiter. An even bigger planet! The 50th brightest star, Alpha Arietis, is fainter than the others but still easy to see in the constellation Aries and currently has one known planet.

Another star with a confirmed exoplanet isn't as bright as the others and doesn't lie near a convenient signpost. Never-the-less, once located, it is easily seen. Situated within the constellation of Eridanus, the river, is Epsilon Eridani. Very similar to the Sun in size and mass, it is the 9th closest star to the Sun and the third nearest naked-eye star. Because of this, long before we knew it had a planet, it was designated as the star orbited by Vulcan, home of Mr Spock. Unfortunately, when we did discover a planet, it was a bit too big to be Vulcan. As disappointing as that is, I hope any beings currently living on it will live long and prosper.

Finally, we have our stellar next-door neighbour, the dwarf star Proxima Centauri. I am cheating a little here because Proxima is not visible, which goes against the original premise of which bright stars have planets. However, at 4.22 light-years away it is the closest star to the Sun, so I think we can give it special dispensation. It is also part of the triple system that makes up Alpha Centauri. Collectively, the other two stars of the system are visible as the third brightest star in the night sky, so we can justify including Proxima if we talk visible star systems that have planets. The planet around Proxima is a super-Earth. With a mass of 1.27 times that of Earth and an estimated radius of 1.08 times Earth's radius, it takes 11.2 days to orbit Proxima and lies a mere 7,250,000 kilometres from its star. This will probably be the first exoplanet humans will visit.

The LMC and SMC

During the Southern Hemisphere summer months, high in the south of an early evening you will find two large hazy patches that look like bits of the Milky Way that have broken off. These patches, however, are not part of our galaxy. They are two of the closest galaxies to our own.

Although known in prehistoric times, it was Ferdinand Magellan on his voyage around the world in 1519 CE that brought them to the attention of the western world and consequently gave them their names, the Large and Small Magellanic Clouds (LMC and SMC for short).

Both galaxies appear to be irregular shaped, and their nearness to our own Milky Way galaxy suggest they are interacting with us and possibly each other. Eventually, all three may even merge to produce one galaxy.

The LMC is about 163,000 light-years from us, while the SMC is about 200,000 light-years away. Measuring the distances to these galaxies was a critical event in the history of astronomy as it then allowed the use of a particular type of star (called a Cepheid variable) to be used as a means of measuring distances in the universe.

Comprised of several billion stars, the LMC is about 1/7 the size of the Milky Way and contains the largest nebula visible in the universe, the Tarantula Nebula. The spidery appearance of the Tarantula Nebula is responsible for its name, but this tarantula is about 1,000 light-years across.

The Large Magellanic Cloud galaxy.

The Small Magellanic Cloud galaxy and the globular cluster 47 Tucana.

If it were located at the distance of the nearest stellar nursery to Earth, the Orion Nebula, it would cover a width of about 60 full moons across the sky. I personally can't see how it looks like a spider, so whoever came up with the name must have had a vivid imagination. In 1987, the closest supernova to the Sun since the invention of the telescope was also seen in the LMC, not far from the Tarantula Nebula.

With only several hundred million stars, the SMC is slightly smaller than the LMC. Sitting beside the SMC, but not part of it, can be found one of the finest globular cluster of stars in the sky, 47 Tucanae. Made up of a few million stars in a volume about 120 light-years across, light from the cluster takes about 13,000 years to reach us.

Exploring for yourself: Discovering galaxies

While outside making all your other observations, have a look at the three galaxies easily visible in the night sky: the Milky Way, the Large Magellanic Cloud, and the Small Magellanic Cloud. These are the largest structures in the universe visible to the unaided eye.

The Milky Way is visible all year round, but the best time to look in the early evening is during winter. It's at this time the centre of our galaxy is directly overhead and, therefore, so are some of the best objects visible. All you need to explore the Milky Way is to lie on your back in a location away from any lights and look up. This was one of my favourite things to do when working in Central Australia and although impossible, at times I could almost convince myself I could see the Milky Way in three dimensions. When exploring the Milky Way, it is possible to see dark dust lanes that split the brighter star path in places, the central bulge of the galaxy around the constellations Scorpius and Sagittarius, and fine structure and details within the Milky Way itself. After exploring the Milky Way with your eyes, use a pair of binoculars to look at some of the finer details. Through binoculars they start to show their true nature as vast clouds of gas and dust (nebulae) and clusters of newly born stars or a ball shape of ancient stars (open and globular star clusters).

The other two galaxies, the LMC and SMC, are best seen in the early evening during the summer months. The following diagram shows their location relative to the Southern Cross. You may have to rotate the map until the Southern Cross matches its orientation in the sky, but once done it will then be possible to locate the two galaxies.

Living in the Northern Hemisphere it is possible to see the small, faint Andromeda Galaxy (the only other galaxy visible to the unaided eye), so when people venture south of the equator they occasionally have trouble locating the SMC and LMC as they don't realise just how big they are. The easiest way to recognise them is to think of the LMC and SMC as looking like bits of broken off Milky Way.

The Southern Milky Way and its dust clouds.

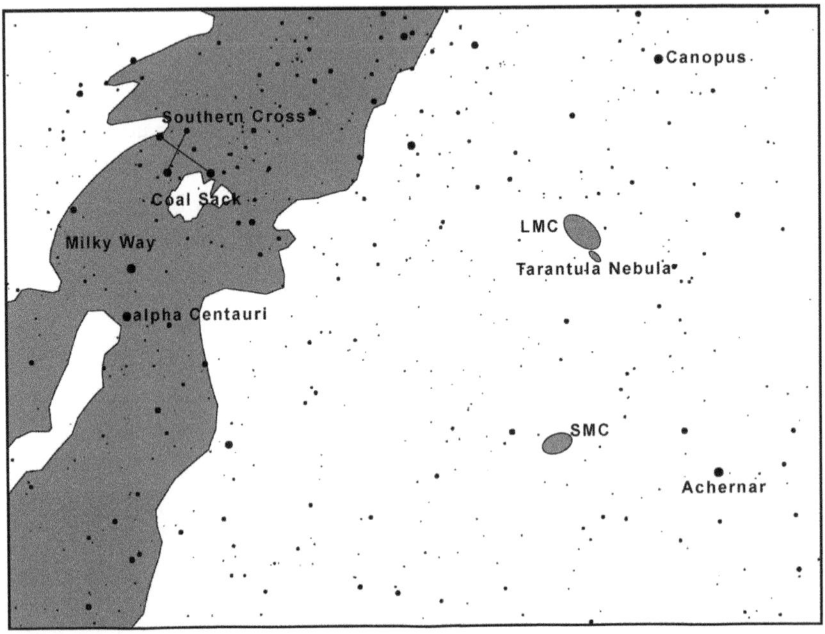

The location of the Large and Small Magellanic Clouds relative to the Southern Cross.

Crux and the Magellanic Clouds.

SN1987A

Roughly 163,000 years ago a star named Sanduleak -69°202 decided it was time to self-destruct. It blew itself apart in what is known as a supernova explosion. Although the explosion took place long ago, the star resided in the Large Magellanic Cloud, so the light didn't reach us until the 24th of February 1987 CE. Interestingly, archival photographs show that the star's brightness was constant for 100 years before its death, and there was no sign of instability.

The supernova, known as SN1987A, was the brightest death of a star seen from the Earth since the supernova recorded by Johannes Kepler in 1604 CE. That star, however, was in our galaxy, a mere 20,000 light-years away.

The supernova was discovered by a couple of astronomers working at a telescope in Chile, but it quickly became bright enough for anyone who knew the sky to notice it. The astronomers were fortunate. They saw the supernova when they took a break and went outside to do something not many professional astronomers do, they looked at the night sky. If they hadn't done that just when they did, they would have lost the discovery to any number of other people.

Simply stated, a supernova occurs when a star's fuel supply driving its fusion reactions has been exhausted. Without the outward pressure generated by the energy at its centre, gravity takes over, and the star rapidly collapses in on itself and then explodes cataclysmically. SN1987A brightened in just three hours, faded and then took almost three months to reach its maximum. It wasn't until May, 80 days after its discovery, that SN1987A attained its peak brightness. At its peak it was easily visible from the centre of Sydney and if you knew where to point a telescope, visible during the daytime.

At the time the supernova occurred, I was working at Sydney Observatory, which lies on the northern side of the city. That meant to see the supernova we had to look over the bright city skyline. Even so, the night after we heard of the supernova, a few of us walked out into the grounds of the observatory, looked south and there it was. We did not have to search for it or use a telescope. It stood out easily in an otherwise empty patch of sky devoid of fainter stars due to the city lights. It was quite exciting and ultimately we were able to see the supernova, with just our eyes, for quite a few months. I hope I get to see another naked-eye supernova in my lifetime.

SN1987A is currently the best studied supernova of all time. Immediately after the announcement of the discovery, virtually every telescope able to see the explosion was turned towards it and observed the explosion with every conceivable instrument.

The Orion Nebula

Stars are made out of vast clouds of gas and dust known as nebulae. Within a cloud, gravity starts to clump the gas and dust together into ever increasing bigger balls. Eventually, these balls of gas reach a size where nuclear fusion begins in their centre, and the stars start to shine.

Typically, there is enough gas and dust in a nebula to make thousands of stars. However, once the first few start to shine, the light and radiation they produce blow away whatever remains of the cloud. This prevents any more stars from forming and a group of a few hundred brand new stars, known as an open cluster, is all that is left.

Due to the complex nature of the original cloud, when the stars form, they are all moving in different directions. Eventually, they will move far enough apart for the tight grouping at their birth to disappear and the stars to scatter into the general stellar population.

The brightest and most easily seen of these stellar nurseries is visible during the southern hemisphere summer in the constellation of Orion and is known as the Orion Nebula. Looking at the middle point of light that makes up the handle of the saucepan (the sword of Orion), with just your eyes will reveal it isn't a star-like point of light, but rather a faint area of fuzziness, the nebula. Binoculars will remove any doubt.

The nebula has enough material to make thousands of stars the size of the Sun. However, it won't make that many, as it will likely form only a few hundred stars before the remaining gas is dispersed. It has already made one of the youngest group of stars known. Called the Trapezium, they are perhaps no more than 300,000 years old and new stars are still forming within the group.

M42, The Great Nebula in Orion.

CHAPTER 19

Unidentified Flying Objects

Before we go any further, I need to get something out of the way. Do I believe in UFOs? Yes. Do I think they're due to alien activity? No. Don't get me wrong. I would love to have an alien spacecraft land in my backyard, although they would have to be tiny aliens in a tiny spacecraft in order to fit. To be the person who has first contact with an extraterrestrial being would be amazing, but sadly I don't think it is likely to happen. Neither do I believe it has happened to anyone else.

There are many reasons why it is improbable for the Earth to have been visited by ET and there are reasonable psychological explanations why people believe they have seen or been abducted by aliens. But for the time being, I have a few questions to ponder over, so let's assume aliens *are* visiting the Earth.

Firstly, humans are a very young technological civilisation, so if aliens are visiting Earth, it is reasonable to assume they would be far more technologically advanced than us and would have nothing to fear from humans. So why would they only harass people on lonely roads or in the middle of the night? It would be no easy task to get here, so why come all this way to do nothing but buzz the locals?

Some people suggest the aliens are studying us, but if they wanted to do that why would they not just take a bunch of people from the one place, similar to what we do when we study animals. After all, the aliens would most likely think of us in the same way. And since we couldn't stop them if we wanted to, they get a lot of people in one go.

Others propose the aliens don't make themselves known because they have something similar to *Star Trek's* Prime Directive which bans interference with a primitive species. If that is the case, when they supposedly abduct people they aren't doing a very good job of observing without interfering.

And why are there so many different varieties of aliens and spacecraft designs? Given how difficult it is to get here, why send a different model of craft each time?

So, until someone provides indisputable proof of an extraterrestrial or their spacecraft, we need to remain healthily sceptical. Getting a piece of the spacecraft might be a big ask, but these days everybody has a phone with a built-in camera, so by now someone should have taken a close-up photo. Admittedly these days it is possible to manufacture pictures through readily available software, but if you had seen an alien up close and taken a photo of it with your phone, surely you would at least offer it up for scrutiny and the chance to be the first person with irrefutable proof that extraterrestrials exist.

Lights in the sky are not proof of alien spacecraft either, no matter how many people see them. Lights could be anything and usually have natural explanations. Also, supposed memories retrieved by hypnosis are unreliable at best and don't constitute proof that the memory is correct.

Irrefutable proof would be aliens landing in the middle of the Sydney Cricket Ground in front of the world's media, stepping out and saying 'Take me to your leader'. Hopefully, they would then give public tours inside their craft, with me first in line.

I do not believe aliens visit the Earth. I do, however, believe there is life everywhere throughout the universe. I just don't think it has been here, yet.

I may not believe in alien visitations, but I do believe in UFOs. The acronym UFO stands for Unidentified Flying Object. Therefore, anything in the sky you do not have an explanation for is a UFO. It does not mean there is no explanation for it, merely that you don't know what it is. I have looked at the sky almost continuously for 45 years and so far there are only two things I've seen that I couldn't explain at the time. However, at no moment did I ever think I was being buzzed by extraterrestrials, simply that I didn't immediately know what they were.

If UFOs aren't aliens, they must be something natural. Unfortunately, most people aren't familiar enough with astronomical objects, nor do they keep up-to-date with scientific events, well enough to identify the things they see. There are, however, some things you can do to help identify any flying object you do see. Rather than immediately jumping to the conclusion it is an alien spacecraft, quite often the answer is not that hard to find.

Identifying something unidentified

In several of the places I have worked we were the first place people called, usually very excitedly, when they saw a UFO. Trying to identify something over the phone when you haven't seen it yourself is tricky, so we soon learnt to ask specific questions to help try and work out what it was. The same questions might come in handy when you see a UFO. So, if you see something odd in the sky, run through the following questions. Even if they don't help you identify what you are looking at, your answers may help someone else.

Where are you?

This gives you a reference point. For example, if you see something towards the south and you live on the northern side of a city, it might suggest you saw something related to the city itself. Or if you live on the coast and see something over the ocean to the east, it might suggest an astronomical reason.

What is the date and time?

When you saw it allows you to check if any events were going on at the time that might have produced what you saw. It could be anything from a nearby festival with an advertising blimp, to a visible rocket launch, or Venus rising.

What was the weather like?

Knowing what the weather was like might help narrow down possibilities. If it was cloudy, you might have seen a searchlight shining onto clouds. If it was clear, then it rules out options such as searchlights.

What did you see?

Write down/record what you saw. Don't leave it until hours or days afterwards. Your memory isn't as good as you think. By recording it straight away, you will be able to keep things straight as you do your research.

What direction was it?

Note what direction you first saw the UFO, where it travelled and where it eventually disappeared. This helps identify possible trajectories.

How long did it last?

Was the UFO event very brief or did it last for a count of 30? The timing doesn't have to be perfect, but it is an indicator of how long it was visible and the length of the sighting can suggest options. Seeing something for a count of 20 is obviously longer than if you saw it for a count of 5, but merely saying it was short or long is subjective and not helpful.

Did it move and if so, how fast?

Did the UFO move extremely fast, or was it slow and even paced across the sky? This might suggest a satellite or meteor. Did it jiggle about erratically in roughly the same position, suggesting it might have been a planet or star twinkling due to atmospheric turbulence.

Did it change colour?

This bit of information can be very beneficial. Whether it was a consistent white or rapidly changing colour can tell you a lot.

How far above the horizon was it?

It is impossible to judge distances in the sky, so giving a height or distance for anything in terms of kilometres is meaningless. There is no point in saying something like 'It was 500 metres off the ground and two kilometres away' as you cannot tell, and by doing so you might be giving misleading information that can confuse an investigation. Instead, provide an estimate of the height above the horizon in terms of degrees or the number of hand spans at arms-length (one hand span equals about 15 degrees in the sky). The horizon is zero degrees, and directly overhead is 90 degrees. Also, because you are not able to determine distance means you cannot tell speed. There is no point saying something like 'it then shot off at 1,000 km/hr' as it is impossible to determine. A more meaningful way to describe its speed is along the lines of 'it went from overhead to the horizon in 7 seconds'.

How big did it appear?

Was the UFO comparable to a star, or was it the size of the Moon? Don't give comparisons to beach balls or cars as, once again, they are meaningless. You cannot tell how far away the UFO is so you cannot describe how big it is. Compare it to something else in the sky, such as a star or the Moon.

How bright was it?

Many people mistakenly equate brightness with size. I have heard 'It was enormous' many times, but on questioning, what they meant was it looked like a bright star. So it wasn't big, only bright. Again, give brightness comparisons to something that doesn't rely on distance. Rather than saying 'It looked like a torch a kilometre away', compare its brightness to a prominent star or the Moon.

Was it seen through binoculars/telescope, filmed or photographed?

Corroborating evidence from sources other than your eyes is always useful. If you have them, look at the UFO with binoculars or a telescope. If you still aren't sure what it is, try and photograph it.

Were there other witnesses and if so how many?

Having more than one person see the UFO at the same time will help you get a complete account when you combine everyone's descriptions.

So, if you do see a UFO and have run through this checklist, what are the most likely explanations? Before you jump to the alien hypothesis, there are many possible natural reasons to consider first, with the most common cases of mistaken identity being astronomical. Unfortunately, most people today do not look up as often as they should, at least not until something grabs their attention.

Unidentified options

Without a doubt, the most commonly mistaken object is Venus, but the following could also apply to Jupiter, Mars and some of the brighter stars.

Many people incorrectly believe that planets don't twinkle. This isn't true, especially if the planet is low on the horizon. Admittedly planets may not twinkle as much as stars, but they do twinkle. Twinkling is caused by the atmosphere, and the more turbulent the atmosphere, the more stars and planets will twinkle. When a star or planet is low to the horizon, you are looking through more of the atmosphere and the twinkling can be particularly bad on some days. The other thing we need to know is that white light is made up of all the colours of the rainbow.

When you have something as bright as Venus low on the horizon, and it is a terrible evening for seeing, Venus will twinkle so much that it flashes through all the colours of the rainbow and appears to jiggle about furiously. This can catch people's attention. One night at work in Sydney I received an anxious phone call about a UFO with flashing lights that was flying around and making radical changes of direction. When the caller used binoculars, the object appeared even more unusual. The person was genuinely concerned and decided to call someone about it. After going through the list of questions mentioned earlier, I looked out my office window and started to suspect it was Venus, which was sitting low on the western horizon at the time. Having an idea on what it might be, I described Venus to the caller and we both agreed it was, in fact, the planet he was observing. He hung up, feeling relieved and perhaps a little embarrassed that he had not recognised such an innocent object as Venus.

Another example of Venus misidentification involves seeing it during the daytime with your eyes. As I mentioned in the section *Exploring for yourself: Venus,* the planet is visible during the day if you know where to look. But finding it without some guide to its location is extremely difficult as your eyes have trouble focusing on a point of light against the bright blue background. However, once you find it, you wonder how you ever missed it in the first place. Now, the number of UFO reports that talk about people watching weather balloons or planes high up that suddenly see a UFO near the balloon/plane which then suddenly disappears sounds remarkably like an accidental Venus sighting. If the balloon/plane is high enough then, to see it, your eyes are effectively focussed at infinity. This is precisely what they have to do to see Venus. If the balloon/plane then happens to pass close by Venus' location, the observer will suddenly notice the bright planet. Look away, however, and your eyes defocus and Venus disappears, making it seem like the 'UFO' suddenly took off at great speed. Sometimes a witness will describe how it circled the balloon. This sounds like a case of perspective producing the effect. If they are locked onto looking at the balloon, then it will appear centred and everything else will seem to move when in fact it may be the balloon moving around due to high altitude winds.

Other typical Venus UFO reports involve being chased for hours in the car. Unlike a tree on the side of the road that appears to flash past as you drive down the highway, the further away an object is, the slower it seems to move past you. Now, even distant trees will eventually disappear out of view, but if you look at something off-world, say the Moon or Venus, you do not see them change position at all. They are always off to the side no matter how far you travel. It almost looks as if they are following you. I've often noticed this effect myself while travelling. Of course, they will slowly move, but this is due to the Earth's rotation rather than your driving skills. Before you laugh, I have had to comment many times to Sydney news outlets on this exact thing in a professional capacity. The media would call for my opinion because they were going to run a story about a family that was followed from Canberra to Sydney by a UFO, or variations on this theme. One time I went to a TV station to look at the footage a family had filmed. As soon as I saw it, I knew it was Venus, and I told them so. That didn't stop them from running the story and somehow my explanation didn't make the cut.

That's not to say I don't make mistakes about UFOs. On one particular occasion, I had just finished a day at work where we had more than the usual number of calls about Venus. It was a Friday and immediately after work, I headed to Culburra, a small town on the coast just south of Sydney to spend the weekend with friends. That night we were walking along the beach when one of my friends asked me what a bright light in the sky was. I could see roughly where he was looking, so, without bothering to look up, I said it was Venus. It was sort of in the right direction. He then asked if I was sure. I said, yes. He said was I sure it wasn't a plane. I said, no, it was Venus, trust me. I still had not looked up at this stage. I had answered questions all day about Venus and I didn't want to go through the same routine with my friends. He pressed me and asked if I was positive it wasn't a plane, to which I replied, I know what I'm talking about. I wanted to change the subject. That ended the conversation until about two minutes later when the plane he had been looking at flew overhead on its way to the local airport. My friends still do not let me forget that night. Nor should they. For them, it was highly amusing, but I learnt a valuable lesson that night, always check something before you put your reputation on the line.

Another commonly mistaken object for UFOs are meteors. Meteors are bits of dust and small rocks from space colliding with the Earth at enormous speeds. As they crash into the atmosphere, they make the air around them extremely hot, melting the meteor and causing a bright streak across the sky before vaporising completely and disappearing. Meteors can be fast or slow, faint or bright, brief or long depending on all manner of things, such as the angle they enter the atmosphere, or whether they are catching up to the Earth in its orbit or running into it. Occasionally a larger meteor, maybe the size of your fist, will collide with the Earth and produce an extra bright display. These are known as fireballs.

As you can imagine, the sudden appearance of a bright light streaking across the sky at enormous speeds then suddenly disappearing could cause concern, especially if it lasts for more than a second or two. I have had many phone calls about spacecraft flying across the sky, displaying aerial acrobatics and then suddenly zooming off so fast they disappear. Sometimes I've been lucky enough to see the same event and realise it was just a meteor.

If you have ever been to the Ayers Rock Resort in Central Australia, you know that even from the centre of the resort, with all its lights, the view of the night sky is truly spectacular. Drive one kilometre out of town, and the view is mind-blowing. While I was working there, conducting tours of the night sky, part of the evening involved pointing out the stars and constellations using just our eyes. We were under one of the wonders of nature, the dark night sky, and the last thing we needed was light, as this wrecks your night vision and upsets the view. One night, I had just finished pointing out a constellation and looked down at the group when suddenly there was an extremely bright flash of light. A second later, there was another. Then another. My first thought was that someone in the group was taking multiple photos using a flash. I was about to ask them to stop when I noticed everyone was looking up, so I did too. What we were witnessing was a rare fireball coming almost straight down at us. As it came through the atmosphere bits of it broke off, and these bits created the bright flashes. We immediately talked about meteors, so everyone knew what they had just seen, but if you didn't know a meteor could do that, what we had seen could have easily been misidentified as an alien UFO.

Meteors can be misconstrued as alien spacecraft and so can satellites and rocket launches. Not long ago, there was excitement where I live because we could see a visible trail left by a rocket launch. We live a long way away from the closest spaceport, so this was a rare event and one that generated numerous enquiries. It was a classic UFO situation. Something dramatic and unknown was due to an uncommon occurrence that took a bit of investigation to identify. In the days before the internet, this event would probably have never been identified.

There are now thousands of satellites in orbit around the Earth. Some can be seen from here on the Earth's surface. They appear as bright, star-like objects moving across the sky at a steady pace and are visible for only about an hour or so after the Sun sets. It may be getting dark down here, but up where they are the Sun is still shining. That means the sky is getting darker, making them easier to see, and they appear brighter as they reflect the sunlight down to us. As the evening progresses, the satellites may not be visible the whole way across the sky. They travel partway across before winking out as they enter the Earth's shadow and no longer reflect sunlight. Sometimes they pulse brightly as they tumble in their orbit. Occasionally they will suddenly flare up as they reflect the light perfectly back to you, returning to normal brightness as the angle between the satellite and your location changes. Lately, there are trains of satellites in orbit. These appear like a string of pearls moving across the sky. If ever there was an alien spacecraft sighting waiting to happen, this is it. Given how many satellites there are in the train, the 'aliens' look like they are swarming, getting ready for an invasion.

While out at the Ayers Rock Resort, we were looking at a crystal clear sky when a slow-moving object appeared and for a count of about 40 (which is an exceptionally long time) travelled across almost the entire sky. This was probably a meteor or piece of space junk skipping off the upper atmosphere before heading back out into space. These are not uncommon, but you have to be lucky to see them. I looked at this particular event as exciting, but I could see how someone else might misinterpret it as a UFO.

So far, I've talked about things I could explain. Is there anything I've seen that I can't explain? Yes, there are two things. Both occurred while I was out in Central Australia. I was conducting a tour when what looked like a small bright cloud moved relatively quickly across the sky. It was a clear night with no other clouds and I surmised the lights of the resort were illuminating it, but the question remained why there was a small 'cloud' moving so quickly overhead and not dissipating. I still don't know. The second event I can't explain looked like the headlights on a plane coming towards us. Nothing unusual there, as planes often have lights shining forward. But this light changed direction and moved away from us while the intensity of the light did not change. Had it been a forward-facing light on a plane, you would expect the light intensity to change. Also, there was no sound associated with it. The lack of sound could be explained if the aircraft was very high, but why the light didn't change is more perplexing. The next day we checked with the airport, and there was no local air traffic so we could discount that. It still, however, could have been a flight passing high over The Centre. I know what a satellite or meteor looks like and it wasn't one of those. To this day, I do not know what it was. My best guess is a plane, but who knows. At no time, however, did I think either of these events were alien spacecraft, only something I had no explanation for at the time.

Special events, such as the Sydney Olympics and the turn of the millennium, can also be particularly fruitful times for UFO sightings. I was working in Sydney during the 18 months leading up to the Olympics, and across the whole period there were special events and more than the usual number of sporting events. A lot of these employed spotlights and advertising blimps to draw attention to themselves. On cloudy nights, the spotlights would reflect off the bottom of the clouds as they moved around and reports of UFOs racing across the sky, making radical direction changes, would spike.

One annoying UFO was an internally illuminated advertising blimp.

It was orange in colour and regularly hovered over sporting events. From a distance, it was impossible to see the advertising on its side, and it just looked like an orange, oval-shaped object stalking Sydney. When it was brought to our attention, we had a look through the telescope and immediately realised what it was. After a few sightings, whenever we saw the blimp in the sky, we would sigh inwardly and brace ourselves for the calls we knew would be coming our way.

Another UFO scare I had to decide on involved a mysterious light in the Sydney suburb of Manly. Two brothers that videoed it claimed it had followed them and that it kept changing shape. One of the TV news had picked up the story and wanted my thoughts to add to what they had filmed. As soon as I looked at it, I knew what it was. One thing that gave it away before I saw the footage was their statement that it followed them. To me, that meant it was probably the planet Venus, and once I saw the images I knew it was. The continually changing shape they said were close-ups of the spacecraft was merely the autofocus on the poor camera trying to latch onto a point source in the sky. The video kept going in and out of focus and the distinct shape displayed when out of focus was simply the internal configuration of the camera. Once I was back at my office, I confirmed Venus would have been visible at the time and in the direction they described. The TV station played the story, with my comment included this time, but the brothers who filmed it still insisted it was an alien spacecraft.

Some people are just naturally predisposed to seeing UFOs as alien spacecraft. Anything they see that isn't obvious to them becomes extraterrestrial in origin, and nothing you say will ever change their minds. Throw in tiredness when driving on country roads, and it's not a big leap for people to transform everyday natural things into UFOs.

Many things foster reports of unknown flying objects. Apart from the above examples, other common possibilities (and this is by no means an exhaustive list) include auroras, sundogs, anomalous atmospheric refraction, weather balloons, reflections from fog and mist, headlights in the distance, eye defects, photo defects, radar scattering and ghost images and, of course, deliberate hoaxes. Unknown does not mean alien. It merely means you don't know what it is, and the process of investigating provide opportunities to discover something new about the world.

Useful unidentified flying objects

I included this section about UFOs because if you spend much time looking at the night sky then at some point, you will see something that seems out of the ordinary.

But apart from that, I think talking about UFOs is an excellent way to excite people about science. It is natural to be fascinated by the prospect of alien life and the possibility it may be visiting us. Admittedly some people get a little carried away, but most have a healthy interest, and that can lead them into a deeper understanding of science. By discussing why and how aliens might get here, you can introduce an audience to Astronomy and Physics. Considering the possibility of extraterrestrial life and where it might be is an excellent way of introducing people to Biology. And exploring possible explanations for UFOs introduces even more Astronomy, as well as some Earth Sciences, Atmospheric Physics and Human Psychology.

To counter the growing and dangerous, anti-science mentality in today's society, we need to increase the general scientific knowledge of the population. Too many decision makers are ignorant of how the world works. Also, the advancement of science depends on public funding and people receiving scientific educations that enable them to make informed decisions. For me, it is a natural fit to use interest in UFOs to introduce them to the wonders that science reveals about the real universe.

Part Three
Telescopes and Stuff

CHAPTER 20

What is a telescope?

Science was revolutionised in September 1608 CE when a Dutch spectacle maker, Hans Lippershey, submitted a patent application for 'spy glasses', which later became the basis for all optical telescopes.

The story of the actual invention is shrouded in myth, but the various versions all follow the same theme. While away one day, Lippershey's assistant was playing with some lenses in the shop. He put one in front of another and saw that things appeared much closer. When Lippershey returned, the assistant showed what he had found to his boss. Lippershey immediately recognised the value of this discovery and developed the idea further by holding the two lenses rigidly in place with a metal tube. He applied for a patent, but it was ultimately turned down as they decided it was too easy to copy. However, the idea was now public knowledge. Lippershey offered his invention to the Dutch government as a secret weapon in their war against Spain. But due to its simplicity, the mere mention of the device was all it took for their enemies to make their own versions, so it wasn't much of a secret.

There were numerous near misses before the eventual invention. Some even claimed to have invented a telescope before Lippershey. Whether they did or didn't is irrelevant as they did nothing with the discovery and it was Lippershey that made the world aware of it. For a long time, most people assumed it was Galileo who had invented the telescope. This misconception probably arose from the fact that Galileo was a lot more famous, not shy of a bit of self-promotion, and it was only a few years after its invention that he first turned the telescope towards the universe and made discoveries that shook the world.

So, what exactly is a telescope?

The term 'telescope' now covers a wide range of instruments, not just one that collects visible light, allowing you to look through it with your eyes. It also applies to any instrument that collects and focuses radio waves, infrared light, microwaves, ultraviolet light, x-rays and even gamma rays. For most of these forms of light, the atmosphere is a big problem by blocking most of the electromagnetic spectrum and not allowing it to reach the surface. To see some frequencies we have to put the telescope into space, which is why we now have space-based telescopes that look at the universe across the entire range of the electromagnetic spectrum. Because the only radiation that does make it to the ground is visible light and radio waves, the two main types of telescopes you are likely to come across are radio telescopes and optical telescopes.

Radio telescopes have to be large since the wavelengths they study are large. But because the wavelengths are long, it does mean the surface of the telescope doesn't have to be solid to reflect the radiation, so the surface is often made out of wire mesh instead. With radio telescopes, we can see any object in the sky that emits radio waves, giving astronomers another view which, when combined with observations from other types of telescopes, helps them to understand the universe better.

As the name suggests, optical telescopes view the universe in visible light and are the classical type of telescope we usually think of when we mention the word. You are likely to come across two different versions of optical telescopes. One is known as a terrestrial telescope, used primarily for things such as bird watching or admiring scenery from a lookout. Binoculars are a typical example of this type. The other type, an astronomical telescope, is used mainly for observing the sky. A terrestrial telescope has an upright image, while in an astronomical telescope the image is upside down. They are essentially the same thing, but a few optical tricks invert the image of an astronomical telescope to create the terrestrial telescope.

One last variety of telescope is a recent invention and struggles to fit into the usual definition. Nevertheless, it is a telescope. For the first time since the instrument made by Lippershey, we have been able to observe the universe in something other than the electromagnetic spectrum. We now have the ability to view the cosmos in gravitational waves (ripples in the fabric of spacetime predicted by Einstein's Theory of General Relativity). At the moment there are just a handful of these exquisitely sensitive observatories around the world, but more are planned to be built.

CHAPTER 21

How do optical telescopes work?

The main aim and principle use of any optical telescope is to make small things that are a long way away look much bigger. But how do they do this?

Everyone is familiar with a magnifying glass. We use them to magnify small printing in a book, get a closer look at a flower and so on. But how does a magnifying glass magnify? Using just your eyes, the closer you bring them to something, the bigger it looks because a larger image is formed on the back of your eye. But, by themselves, our eyes are unable to focus on anything closer than about 10 centimetres, as the lens in your eye just can't deform any more to bring things into focus. But if we could bring our eyes closer and still focus we would see the object even bigger. Effectively this is what the magnifying glass allows us to do.

A magnifying glass is just a lens with a handle. When light from something in the distance passes through a lens, it is bent so that it all meets at one point on the other side, known as the focal point. Since parallel rays of light can enter the lens from slightly different angles, an image is formed not just at one point but on a focal plane. The distance from the lens to this focal plane is called the focal length.

But if we go the other way around and, for example, look at some

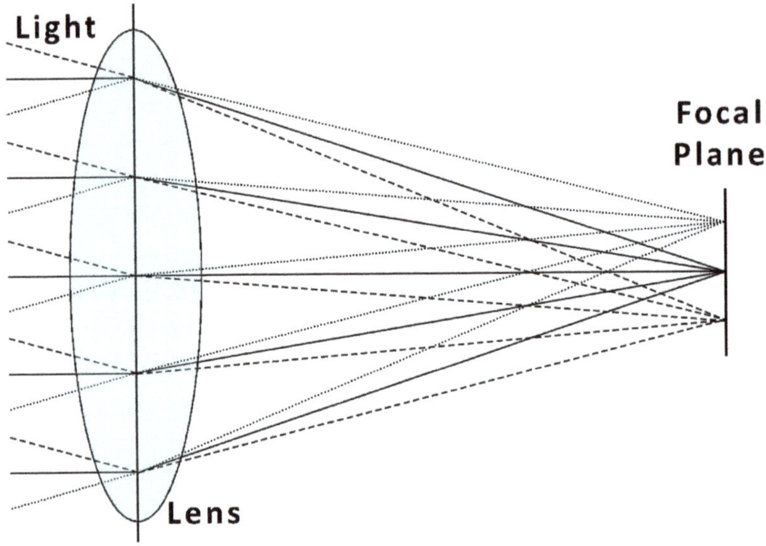

A single biconvex lens focusses light passing through it to a focal plane.

small printing in a book through a magnifying glass, the first thing we do is move the lens in or out until the printing is in focus. This puts the printing at the focal plane of the lens. Light from the printing comes up off the page and passes through the lens. Since different parts of the printing are at slightly different positions, the light rays from each point pass through the lens, and your eye then generates a virtual image of the printing. This virtual image appears larger than the original, and consequently, the printing appears magnified.

A magnifying glass is not bad at magnifying things, but it has one major flaw, you have to be very close to the object, no more than the focal length of the lens away. This doesn't help when we want to look at distant birds in a tree, or the stars. If we could get close enough to the stars to use a magnifying glass, we wouldn't need the magnifying glass in the first place. It is still possible to use a magnifying glass to see the stars, but we need to do a bit of trickery first. If we can't get close to the stars, what we need to do is bring the stars close to us. As we now know, when light passes

When viewed at the focal point of a lens the image appears upside down.

through a lens, it is focused to a point and makes an image on the focal plane, effectively making a large area small. If we point it at the stars an image of that patch of sky is produced, and we have brought the stars to us. If we then pass the light from this image through another lens (called an eyepiece in a telescope), we get a virtual image of the stars and they appear much bigger. In other words, we have magnified them.

Of course, if you use a lens with the same focal length as the one making the image then there won't be any magnification, you will see the image the same size as you would using just your eyes. Similarly, if the eyepiece lens has a longer focal length than the one making the image, then everything will look smaller. To see this effect, look through a telescope (or binoculars) the wrong way, essentially swapping the order of the lenses, so the longer focal length lens is closest to your eye, and you will see everything appears smaller. To magnify the image, the focal length of the lens you are using as a magnifying glass has to be shorter than the focal length of the lens you are using to make the image. The more significant

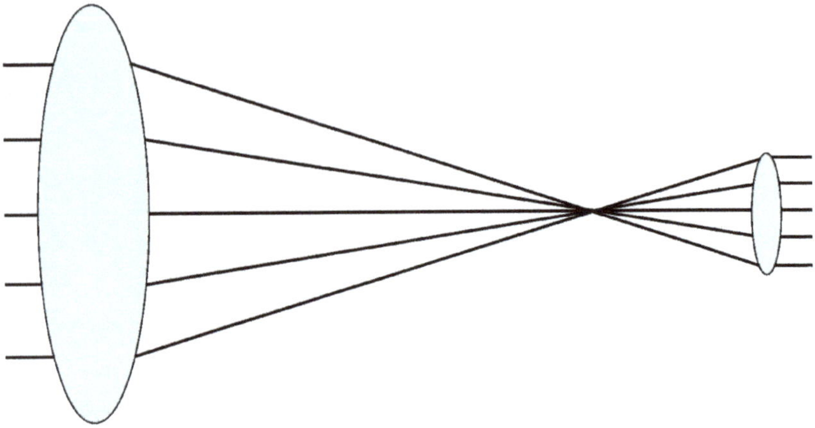

The front, primary lens brings the light to a point, and the second lens magnifies the generated image.

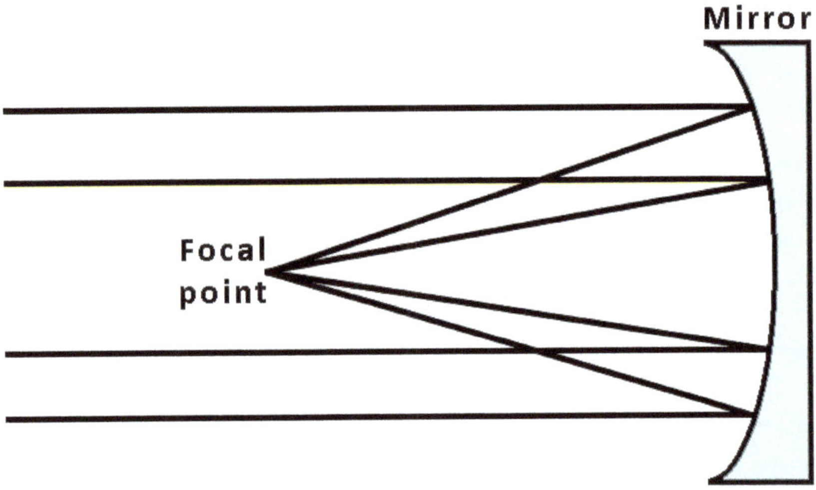

A parabolic mirror does the same job as a biconvex lens and focusses light that strikes it to a focal point.

the difference, the more the image magnifies. What we have just described is a telescope (more specifically, a refracting telescope), an instrument that will allow us to see objects a long way away much closer.

It is also possible to use a specially shaped mirror instead of the first lens to focus the light and make an image. These telescopes are called reflecting telescopes (because they reflect the light).

Incidentally, all astronomical telescopes turn the image upside down. The simplest way of making a telescope does this naturally, similar to the way your eyes work. To see an upright image, you need to put in extra lenses to turn the image the right way up. But seeing things upside down in space doesn't matter. After all, something that is the right way up to us is upside down to someone in the other hemisphere anyway. If we can't agree on which way is up and which way is down using only our eyes, why does it matter when looking through a telescope? So astronomical telescopes keep it simple (and upside down) to save on cost and image quality. But for telescopes specifically designed to look at things on the ground, an effort is made to have the image the right way up.

To sum up: a telescope is a straightforward instrument using a lens or mirror to focus the stars and make an image, and another lens to magnify the image at this point. Your eyes are, in essence, small optical telescopes. The main difference is that instead of a second lens allowing you to see the image created by the front lens, the image is put directly onto the back of your eye and your brain then does all the work.

Interestingly, the brain is such a fabulous piece of evolutionary machinery that if you were to wear glasses that made everything look upside down, after feeling nauseous for a couple of weeks your brain would adapt and everything would look the right way up again. Take the glasses off, and you would have to repeat the process once again until your brain adapts and things look the way they should using just your eyes.

There's one last thing to say about eyes and telescopes. Since your eyes are small optical telescopes, it means that if you cannot see through something with your eyes then no optical telescope, no matter how big, can see through it either. Admittedly if the clouds are thin and you can't quite make something out with your eyes a telescope might help, but it doesn't make the clouds disappear, it only shows the thin cloud covered object slightly better. If the clouds are thick enough that you cannot see through them with your eyes, a telescope cannot see through them either.

Whenever I talk about telescopes, a few basic terms keep cropping up. Understanding these terms will help you get the best out of your telescope.

CHAPTER 22

Telescope bits and pieces

Aperture

The aperture of a telescope is the size or width of the primary lens or mirror and determines how much light the telescope collects. This is the most crucial aspect of a telescope. Since things in the night sky are very faint, it is essential that the telescope collects as much light as it can. For astronomical purposes, bigger means better (assuming the quality is good) which means the larger the aperture, the more light from the object it gathers and hence the brighter the image. Not only does this make it easier to see a bright object, but it also means the fainter an object can be and still be visible. The relationship is with the area of the lens or mirror, so if you double the aperture, you quadruple the light gathering capacity of the telescope. It also means that the resolution is better. The technical definition of resolution is 'the smaller the angle between two point objects and still be able to produce distinct images', but in essence, it means the finer the detail you can see. As an example, an average-sized amateur telescope has an aperture of 20 centimetres. Theoretically (assuming conditions are perfect) this will allow details to be seen as small as 1,160 metres across on the Moon (384,000 kilometres away). One of the largest telescopes in the world, the Keck telescope in Hawaii, is 10 metres across and can theoretically see details just 25 metres wide on the Moon.

Focal length

The focal length of a telescope is the distance from the primary lens or mirror to the point of focus near the eyepiece. In itself, this doesn't mean much, but it does determine some other specifications.

A 280 mm telescope and an 80 mm telescope.

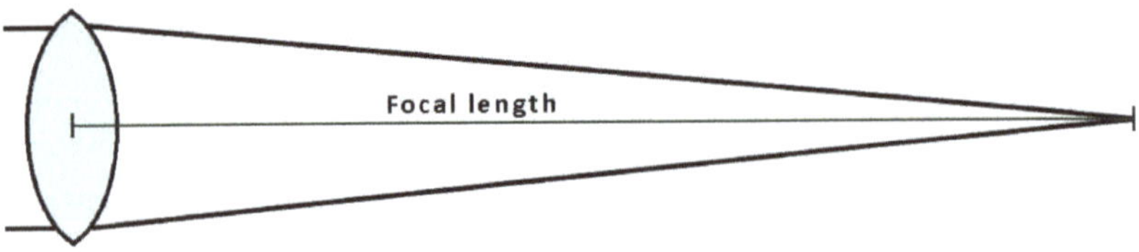

The focal length of a lens is the distance from the lens to the focal point.

Magnification

Magnification is often misunderstood and overrated. The magnification of a telescope is not its most important aspect and can be changed quite easily. To work out how much magnification you have, divide the focal length of the primary lens or mirror (usually found written on the side of the telescope) by the focal length of the eyepiece being used (traditionally found written on the side of the eyepiece). Since you can't change the primary lens or mirror, to change the magnification, all you have to do is change the eyepiece. It is a trap to think high magnification means high performance. Too much magnification is worse than too little. The more you have, the fainter and more blurry the image becomes. Fainter, because with a bigger image, what light is collected by the telescope has been spread over a larger area. Blurrier, because you are also magnifying any distortion caused by the atmosphere.

Focal ratio (f ratio)

The focal ratio is related to the focal length of the primary lens or mirror. It is the focal length divided by the aperture. For example, a telescope with a focal length of 1,400 millimetres and an aperture of 200 millimetres has an f-ratio of seven. The f-ratio is the same as the f-stop you find on a camera, only with a telescope it is not so easily changed. The higher the focal ratio (greater than f10), the smaller the patch of sky the telescope looks at, which makes them excellent for planetary work and splitting double stars. Shorter focal ratios (less than f6) tend towards wider fields of view and lower magnification, which make them best suited for nebulae and galaxies. In between focal ratios make for good, all-round, general purpose telescopes.

Refractor (Refracting telescope)

A refracting telescope uses a lens at the front of a tube to gather and focus the incoming light down the tube to another lens, the eyepiece. There aren't many large refractors in the world, and the largest is not very big by today's standards. Refractors tend to be small and expensive as they require the material the lens is made from to be perfect all the way through. They also require both sides of the lens be ground to a perfect shape. In modem refractors, the primary lens is often made up of two or three segments, each requiring each side to be perfectly ground. All of these add up to lots of time, effort and cost to make.

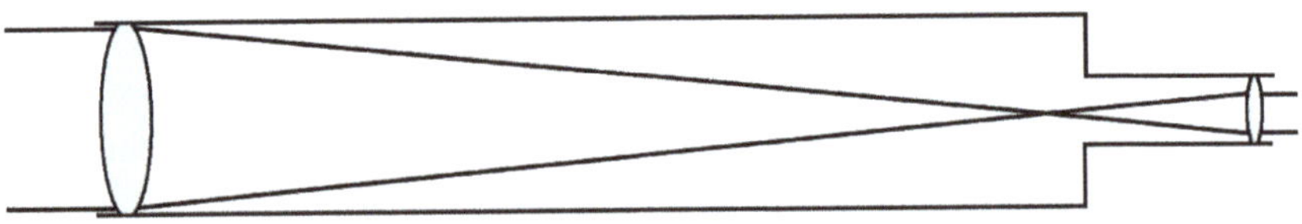

A refracting telescope has one lens at the front of a tube and a second lens to look through at the other end.

An 80mm refracting telescope.

Reflector (Reflecting telescope)

A reflecting telescope uses a mirror instead of a lens to focus the light. The mirror does the same job as a lens but in a slightly different fashion. The original style is the Newtonian, where the primary concave mirror focuses the light back up the tube to a flat mirror (the secondary mirror) which bounces the light out the side of the tube to the eyepiece. This style of telescope is a favourite of the amateur astronomer. Given that only the surface of the mirror needs to be perfect and not the rest of the material, plus only one surface needs to be ground, reflectors tend to be cheaper than equally sized refractors. Also, from an engineering point of view, a mirror is easier to support, and hence it is easier to build much larger reflecting telescopes than refractors.

Catadioptric telescope

This style of telescope uses both lenses and mirrors. The primary example is the Schmidt Cassegrain Telescope. A concave mirror focuses the light back up the tube, but this time a convex mirror (the secondary mirror) bounces the light back down the tube, through a hole in the main mirror to the eyepiece at the back. However, as the light initially enters the tube, it passes through a lens (called a corrector plate) that eliminates an optical problem with the primary mirror known as spherical aberration. This style is also a favourite of amateur astronomers. Schmidt Cassegrains are more expensive than other reflecting telescopes, although still not as expensive as equally sized refractors. Needing a corrector plate at the front as well as a mirror means they tend to only come in a limited range of sizes. However, they are a nice, compact design.

A Newtonian telescope on a Dobsonian mount.

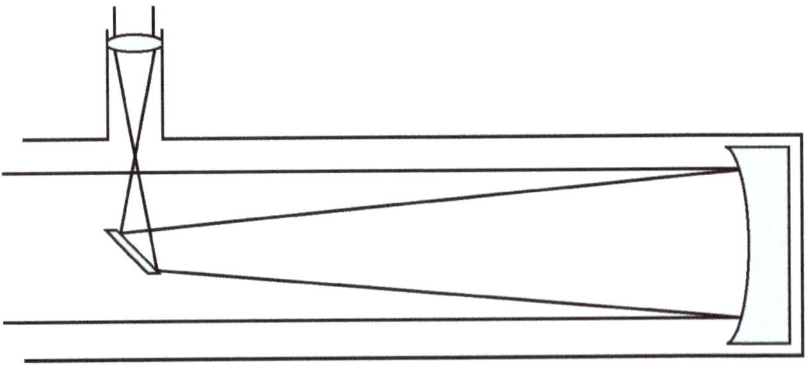

A Newtonian telescope uses a parabolic mirror at the bottom to focus the light back up the tube before reflecting it out to the side and a lens for viewing.

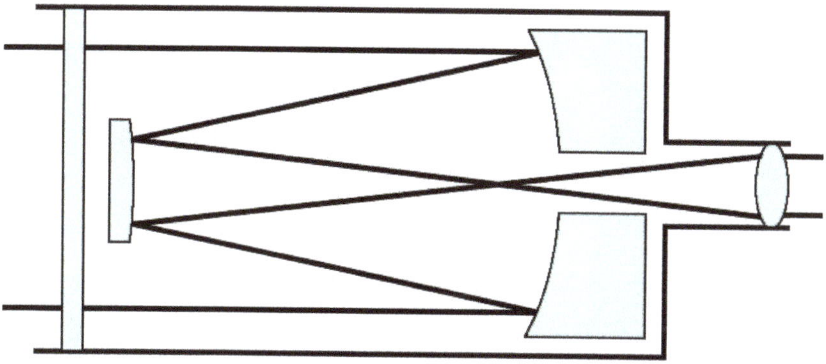

The light in an SCT first passes through a correcting lens to adjust for using a spherical mirror instead of a parabolic mirror.

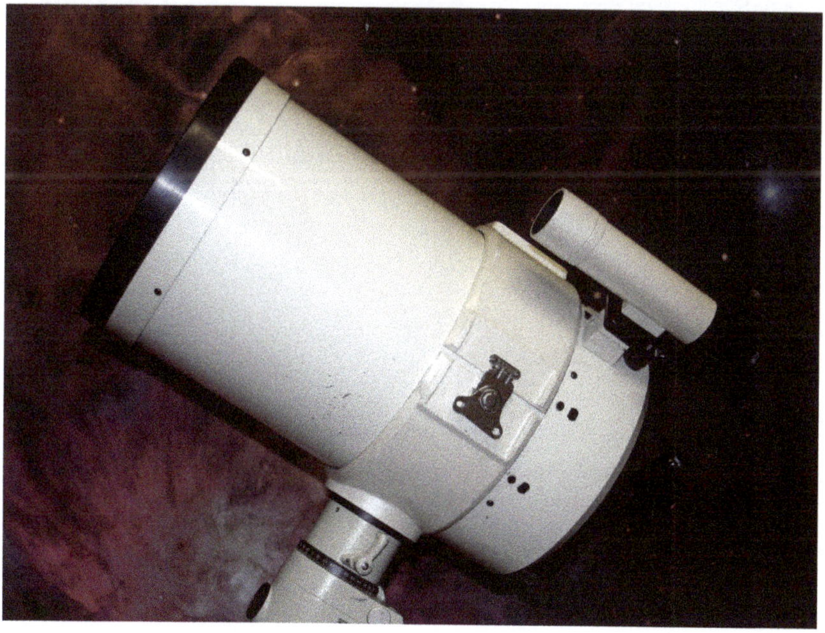

A Schmidt Cassegrain telescope (SCT) is shorter than a Newtonian telescope but still uses a mirror to focus the light.

The Dobsonian is a variety of Alt-Azimuth mount where it swings the telescope up and down, left and right.

Alt-azimuth mount

Alt-azimuth stands for altitude (height above the horizon) and azimuth (the angle measured in an easterly direction from true north). This system is the cheapest and easiest to use, as the telescope simply swings up and down and from side to side.

Equatorial mount

When we look at the stars throughout the night, it appears they are all moving, and the Earth is stationary. In reality, of course, it is the other way around. The stars are not the ones moving. It is the Earth spinning below them. However, having evolved living on the Earth means we can't tell this.

Consequently, since it is the rotation of the Earth that causes the motion of the stars, as we look at the sky the stars appear to make circles about a particular point, directly above the Earth's rotation pole. If we could align a telescope mount so that one of its axes pointed at this pole, the axis would then be parallel with the Earth's rotation axis. To then follow a star all you have to do is turn the mount about this axis at the same speed as the Earth's rotation but in the opposite direction. This is what an equatorial mount does. Whereas an alt-azimuth mount needs to be moved in both axes to follow a star, the equatorial mount has to move around only one. Attaching a motor then becomes easy, and the object appears motionless in the eyepiece. The biggest challenge is aligning the mount to the Celestial Pole in the first place. There are two main versions of the equatorial mount. The German Equatorial Mount has the telescope on one side of the axis that points at the Celestial Pole balanced by counterweights on the other side. The Equatorial Fork Mount has the telescope sitting between the tines of a large fork arrangement aligned along the mount's axis pointing at the Celestial Pole.

Eyepiece

Eyepieces are the most crucial accessory for any telescope. They are the small removable lens you look through at the rear of a telescope. Without them, you do not have a telescope. Eyepieces come in a range of focal lengths, diameters, optical designs and quality. A poor quality eyepiece can ruin a telescope, so it is imperative to use good quality ones.

The German Equatorial mount swings the telescope around an axis aligned to the South Celestial Pole.

Different sizes and styles of eyepieces.

Finderscope

Since the main telescope is looking at only a tiny patch of sky, it can be challenging to find what you want to observe. To overcome this problem, a smaller telescope with less magnification and a set of crosshairs is mounted on the side of the main telescope. Providing they have been correctly aligned, the finderscope allows you to find what you want to look at and then you can see it in more detail through the main telescope.

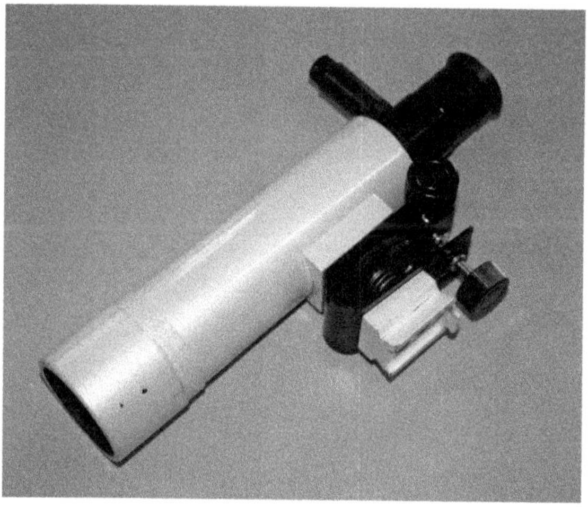

A finderscope is used to point the telescope at the observed object.

CHAPTER 23

Buying a telescope

Beginners in astronomy often ask 'Which telescope should I buy?'. Unfortunately, the short answer is that there is no best beginner's telescope. There is no one style or brand of telescope that is obviously for someone just starting. This lack of a clear-cut answer is confusing for the person wishing to buy a telescope, but fortunately, there are a few simple things you can do to help decide which one is most suitable for you.

The very first thing you need to ask yourself is just how interested in astronomy are you? Telescopes and related equipment can be expensive, so you need to decide how committed you are before outlaying potentially large sums of money. It is not uncommon to hear stories of people spending thousands of dollars on a telescope only to use it for a month then store it away for years before selling it after a house cleaning spree.

Secondly, with so many different types of telescopes, mounts and accessories available, there is a temptation to get carried away and buy the best of everything. However, a lesson quickly learnt is that without any trouble at all, you can spend anything between a few hundred dollars to a few hundred thousand dollars on a telescope. You need to decide just how much money you are prepared to outlay. In other words, how far will your budget stretch? Once you have decided this, you can start looking at what will best meet your needs within your budget.

Ultimately, the answer to which telescope you should buy is the one that will get used the most. It doesn't matter how much the telescope costs, how big it is, or how good the quality is, if you use it and get enjoyment from it then it is the ideal telescope for you. But, keeping in mind that if it is so cheap and nasty that you quickly get frustrated with it, or if it is the highest quality telescope but so big you never set it up, then it has been the worst telescope.

With this in mind, if your budget is small, or you're not sure how committed you are to astronomy, it might be a good idea to explore the wonderful world of binoculars. They are easily transportable and easy to use. And if your interest wanes, they are useful for other things and you haven't outlaid an enormous amount of money. The downside is that you won't see as much with binoculars as you will with a telescope.

If you decide you want to own a telescope, then there are plenty of good options out there for a reasonable price. But before you head off to the store, there is another option you might like to consider if you are reasonably handy at building things. I am talking about the alternative of making your own telescope. Contrary to what you might expect, it isn't that hard to build your telescope and make a reasonable job of it. With a bit of patience and the following of instructions, you can grind your own mirror and assemble the telescope from its components. Building the telescope yourself has the advantage of making the completed unit slightly cheaper than an off-the-shelf telescope. That means you can generally afford to make a bigger one than you could buy ready-made. Plus, it is a great experience that allows you to more fully understand how a telescope works. The disadvantage is that although perfectly functional, they tend to be more simple than a store bought telescope. It won't necessarily affect the optical performance, but since most people aren't engineers, it will affect the general finish, especially the mount, and may limit how you can use your instrument. If you are prepared to put up with these limitations, or perhaps would like a big telescope but only have a small budget, then making your own telescope is a satisfying option and an adventure.

I made a telescope myself when I was a teenager. As I was growing up, I was obsessed with looking at the stars and finding out everything I could about them. I used to regularly annoy my parents to buy a telescope for me, but I was one of five children and my parents never had much spare money for something like a telescope. After years of pestering, my parents must have realised that my interest wasn't going to go away. So, in a moment of genius, for my 16th birthday they did the next best thing to buying a telescope, they bought me the mirror blanks and accessories to grind my own 20 centimetre diameter mirror. Grinding a mirror was a new experience for all of us. No one knew what to do or what to expect, but I had a book that explained everything so, as only a teenager can do, I figured it couldn't be that hard.

A telescope mirror is ground by rubbing two thick flat pieces of special glass over each other. The bottom piece of glass is fixed to a table and you rub the centre of the top glass over the outer part of the bottom glass. In between the two is a coarse grit that wears the glass away. After a little while, you turn everything a bit and repeat, this time rubbing different areas of both bits of glass. After doing this for hours and hours and hours, you change to a finer grit and repeat, for hours and hours and hours. You then switch to an even finer grit and repeat the process yet again. Teenagers aren't renown for their patience and I have to admit that at times I struggled to keep going. But slowly and surely the outside of the bottom glass gets worn away, creating a convex shape, and the centre of the top piece is worn away to create a concave surface. As the grit gets finer, the surfaces become smoother until eventually, you have concave and convex pieces of the glass that are incredibly smooth. You then throw away the convex one (unless you need a paperweight) as it is now useless. You keep the concave glass as it is now your telescope mirror. It still needs a reflective coating applied to turn it into a proper mirror, but essentially the mirror has been made.

The mirror is, of course, just one part of a telescope. I still needed to buy the tube and other bits and pieces. It took a while to save the money from part-time jobs, but eventually, I had all the necessary bits to complete the telescope. I painted the tube, drilled holes, measured focal lengths and finally put together my first telescope. I next turned my attention to making the mount. The mount wasn't pretty when I finished, and it wasn't a great piece of engineering, but it worked, and it allowed me to use my very own telescope I had built from scratch. The whole process was such a satisfying experience that 40 plus years later, I still have that telescope. It works, but I haven't used it for quite a few years. The homemade mount fell apart a long time ago, and since then, I have had several commercially made telescopes, so I don't have any need for it anymore. But I can't bear to get rid of it, so it now sits in the sunroom as a piece of artwork in the corner.

If you decide to go ahead and purchase a telescope rather than binoculars or making your own, the first thing to remember when you enter a store (and this goes for online stores as well) is just that, you are entering a shop. That means the salesperson (or website) will be trying to sell you a telescope. But, as with anything you buy, if you know what you are after, you shouldn't have any problems.

I recommend buying your telescope from a store that specialises in telescopes and telescope accessories. Department stores and the like may have a similar telescope and could even be slightly cheaper (although not necessarily), but they also generally lack detailed knowledge about what they're selling. If you have any questions or need after-sales advice, you may find help hard to come by. However, if you buy from a store that specialises in telescopes you will get any professional help you may need. They will also have a more extensive range to choose from.

Once you know how much you can afford to spend, you can now look for the telescope that best fits your needs for this price. But be careful, it is easy to get carried away and want the best of everything when, in order to fit into your price range, many compromises may have to be made. Deciding on the best telescope to buy is a tough decision, as there is no perfect telescope, so the best solution is to have a look at what choices there are within your price range and consider the following points.

What do you mainly want to use the telescope for? Is a refractor or reflector better suited for this purpose?

If you are mainly after faint objects such as galaxies and nebulae, then the larger the aperture, the better. Reflecting telescopes will give you more aperture for your dollar, but they pose other problems such as bulk. Refractors are great for planets, the Moon and double stars. If you have no specific aims, then Schmidt Cassegrains are good general-purpose telescopes. Think about which style will best suit your needs.

A proud owner and his first homemade telescope.

My first telescope 40 years later.

Should you get a computer-controlled telescope or not?

A lot of telescopes are now computer-controlled (often referred to as GoTo telescopes). This allows you to enter the object you want to look at into the computer and the telescope will automatically find it for you. Having the GoTo capability certainly makes finding objects simpler, but does require a bit of fiddling to set up and costs more than a regular non-GoTo mount. If cost is not a big issue then they are handy to have, but not essential. If you do not have a GoTo mount it just means you have to learn your way around the night sky a lot better, which is not necessarily a bad thing. One big advantage of having a GoTo mount is it makes tracking an object easier.

Make sure the telescope has an adequate aperture for what you want to do with it.

The bigger the aperture (wider), the more light it collects and that means the fainter the object can be. It also means the finer the detail (better resolution) visible. Since you are dealing with faint objects in the night sky, there are minimum recommended sizes for an astronomical telescope. The usually quoted minimum aperture for a refracting telescope is 50 millimetres and for a reflecting telescope, 100 millimetres. But telescopes have come down in price over the last decade, so it is relatively inexpensive to start with a larger telescope than these minimum sizes. So, think about how much aperture you want. It is easier to buy a larger telescope first-up than deciding on a smaller one and cursing yourself six months later for not getting more aperture. Some people say bigger is better and advise you to get as big as you can afford, but aperture is not the only thing to consider.

Unless the telescope will be permanently mounted somewhere, are you going to be able to carry it by yourself and will it fit into the car?

Since you may be the only one who ever carries the telescope, how big and how heavy are you willing to lift? Even if you can lift a big telescope, how much trouble is it to do so? If it is a lot of work to carry and set up the telescope, you will probably use it only rarely. A smaller telescope might get used a lot more. And how big is your car? There's no point in having a large telescope if it won't fit in the car. Think maximum practical size and portability.

Does the telescope have good optical quality?

Right across the field of view, stars should focus to a point rather than blurry blobs. As it is impossible to test the telescope on the night sky before buying, you need to make sure you can return the telescope or have it fixed or replaced if it is unsatisfactory. Generally speaking, bad optical quality can be expected if the rest of the telescope is of poor quality. If it looks cheap and nasty, it probably is. Fortunately, most stores that specifically sell telescopes tend to sell quality equipment, so this is not a big problem, even for telescopes at the cheaper end of the market.

Carrying a large telescope any distance is difficult. I have lots of problems moving this one around, especially to remote locations like desert paths.

Does the telescope have a solid mount?

A stable mount makes using the telescope easier and avoids frustration. The mount's movements should be smooth and controllable. If at all possible, avoid flimsy mounts. Think of solidity, stability and quality. If you believe it has all of these, then chances are it has. One of the biggest problems with the cheaper telescope is that to make them affordable, something has been sacrificed. In most cases, it is the quality of the mount, but this doesn't mean the telescope is no good. It just means that to get a cheap telescope, the mount may not the best quality, and you have to work around it. If your budget is relatively unconstrained and you are buying the telescope and mount separately, a good rule of thumb is to spend approximately the same amount of money on the mount as you do on the telescope. This tends to ensure that you have a mount worthy of the expensive telescope you just bought.

The telescope should have a decent finderscope.

Some cheaper telescopes also save costs on the finder, and the ones provided are tough, if not impossible, to use. Once again, if you accept that this is the case, you can work around it.

Get a good eyepiece.

For many telescopes at the cheaper end of the market, a range of eyepieces will be provided. But as the price of the telescope increases, so do the options of having interchangeable eyepieces. These give you greater flexibility when it comes to magnification and field of view. To get the most out of any telescope, you will need a range of eyepieces. This is so you can have a range of magnifications to cover the different sizes and brightnesses of objects. Eyepieces come in a wide variety of designs and quality. A poor quality eyepiece will make your telescope next to useless, no matter how good the telescope itself may be. So, even though a good eyepiece might be more expensive than others, it is well worth the extra expense.

Do not be convinced into buying a telescope based on how much it can magnify.

The magnification of any telescope depends on what eyepiece is used. Plus, the maximum magnification for a telescope under ideal conditions (which rarely happens) is about 100 times for every five centimetres of aperture. There may not be anything wrong with a telescope that advertises its magnifying power but buy on quality or price, not magnification.

If this is going to be your first telescope, then make sure it has clear instructions on how to set it up and use it.

This is especially the case for computer-controlled GoTo telescopes.

And some final thoughts. If it looks complicated to use, it probably is. If it looks cheap and nasty, it probably is. And any telescope will give a better view than the human eye. So, no matter what the quality or cost, if it gives you pleasure and inspires you to go on to bigger and better things then no matter what anybody else says about it, it has been a great telescope.

CHAPTER 24
Using a Telescope

Getting the most from your telescope

Unfortunately, too many people acquire a telescope and then assume using it and finding things in the sky will be straightforward. It never is. To get the most out of your telescope and enjoy the wonders it can show, there are a few things you need to know and do first.

When you get your new toy home, the first thing to do is play with it. See how to put it together and what each of the bits and pieces do. Find out how it moves and how to lock the axes, if you can. This will give you a feel for how to use it, especially with a GoTo mount. Preferably do this inside rather than trying to work out how to put it together and use while standing outside in the dark and the cold. It is also vitally important to make sure you know how to align the optics properly. Known as collimation, the different optical elements all have to be perfectly aligned in order to have the stars appear as points of light rather than fuzzy blobs.

Next, it is critical to align the finderscope before you can find anything. Surprisingly, telescopes rarely come with instructions on how to do this, or people ignore any instructions they do have. If the finder is not aligned, you won't be able to find anything in the sky and just end up getting frustrated. To align the finderscope look through the main telescope at a tree (or house, or cow, or whatever) visible on the horizon, or as far away as possible. Centre the object in the view. Now carefully, and without moving the telescope, adjust the finderscope so that its view is centred on the same thing. The first time you use the telescope to look at the stars repeat the process with a star. This will ensure that the telescope and the finderscope are now precisely aligned on the same spot in the sky. Once you have done this, be careful not to bump the finderscope or you will have to repeat the whole exercise. Aligning the finder is still necessary even if you have a GoTo mount. As part of the setup process you need to be able to correct the computer if required so it can adjust itself.

Most people now come unstuck as they immediately try to find something in the sky. Unless you have a computer-controlled mount that will find things for you, the only objects you are likely to locate are the Moon and a lot of uninteresting stars. Lots of people have told me that they have a telescope in their cupboard which they bought years ago and all they could find was the Moon (which is almost impossible to miss), so they put their telescope away, never to be used again. Now they want to sell it and do I know anybody who might be interested. When I question them about the telescope itself, some have an instrument that must have cost them a lot of money. But because they didn't know how to use it, their telescope was virtually useless, as they couldn't find anything other than the obvious.

So, to get the most out of your viewing sessions, you need to learn how to find your way around the night sky. If you can't point your finger at the spot where an interesting object is located, you won't be able to point the telescope at it. You can use star maps, apps on your phone/tablet, or computer programs to learn your way around the sky, but all of them require learning the skill of translating what you see on the map/app/program to what you see in the real sky, and that takes practice. Once you have mastered this art and you know how to use your telescope, you will get much more enjoyment from it.

Pointing the telescope

When you can point your finger at where you want to aim the telescope, and the finderscope has been properly aligned, it is now a simple (well, relatively simple) matter to point the telescope itself.

After roughly orienting the telescope to the spot in the sky you want, look through the finderscope and move the telescope up/down/left/right until the object, or at least a very nearby star, comes into view. Centre it in the middle of the finderscope crosshairs. Without bumping the telescope, lock the axes into place if you can and look through the main telescope eyepiece. Hopefully, the object is in view. If needed, centre the object by slowly and carefully moving the telescope. If it isn't, go back to the finderscope stage. And remember, the view in the eyepiece will be upside down and back-to-front. So if you think you need to move the object up and to the right you will have to physically move the telescope down and to the left. You will get used to it after a while. Focus the view by carefully turning the focus knob if need be. And, of course, enjoy.

A colossal warning: The above sounds easy, but mastering the use of a telescope is not straight forward. Like all worthwhile things, using a telescope takes practice. Don't be disappointed if you can't do it straight away. Even the most experienced and adept telescope user had to go through the same learning process. Persevere until you master this art and it will be well worth the time and trouble.

Looking through a telescope

When pointing a telescope at an object, you may find that you can't see a lot of detail at first. Don't panic. Like everything else to do with telescopes, seeing the most detail takes practice. The more you look through a telescope, the greater the detail you will pick out as your eyes and brain get used to it.

The very first thing you need to know about looking through a telescope is that once it is pointing at an object, do not touch the telescope anywhere. Remember that the telescope is looking at a tiny patch of sky, so any vibration will make the view shake. If the view shakes, everything looks blurry. No matter how careful you are, if you touch the telescope, even your heartbeat is enough to make the view move around.

Next, close one eye or put your hand over one eye, it doesn't matter which, and put your eye up to the eyepiece so it is almost, but not quite, touching the eyepiece. Too far away from the eyepiece is just as bad as too close. Once you have the right distance, relax and look straight into the eyepiece. Take your time and let your eye get used to looking. The longer you look, the more you will see as your eye becomes acquainted with the view.

When you are trying to look at a particularly faint object, don't look straight at it, look to the side instead. You will see more detail out of the corner of your eye. This is called averted vision and also takes practice to master. Averted vision works because of the way the human eye is constructed. The centre of the eye deals primarily with colour while the outside of the eye deals with contrast. So, when you look straight at something, you are using the relatively few colour receptors in your eye and the object may be too faint to register. Looking just to the side and using the outside part of your eye means you are using the far more numerous contrast receptors.

Since distances in the universe are so immense, it even takes light time to traverse them. This means that when you look at anything in the sky, you are not seeing it the way it is now. You see the object the way it was when the light, that is only now reaching your eye, left the object. With even the closest star to the Sun, Alpha Centauri, we only ever see how it was a bit over four years ago. This means that the further away an object is, the further back in time you see it. There is every chance that some of the stars you look at have long since died, but the light from their death has yet to reach us. For instance, we saw supernova SN1987A, which we mentioned earlier, explode in 1987 CE when, in reality, the star blew itself apart around 163,000 years ago. It just took that long for the light of the explosion to get to us.

Trying to keep all these time adjustments due to distances and the order that events happened in your head just gives you a headache. We, therefore, tend to think of timelines purely in terms of when we see them here on Earth, not when they actually occurred. As a case in point, the supernova witnessed in 1054 CE, which created what we now see as the Crab Supernova Remnant, is estimated to be around 6,500 light-years away. That means it occurred almost 160,000 years after SN1987A, even though we saw it 900 years before we saw SN1987A.

One of the persistent problems for astronomers and their use of telescopes is the atmosphere. Although essential for humans to breathe, it causes nothing but problems for our view of the stars. Astronomer's call the effect the atmosphere has on the view of the heavens 'seeing'. Consequently, you may hear someone who owns a telescope talk about what the 'seeing' was like last night. To the general, non-telescope owning, public, the effect is referred to as 'twinkling' when using just your eyes to look at the stars. The more a star twinkles, the worse the seeing. The less it twinkles, the better the seeing. To get the best views, pick a night when the seeing is not too bad. This rapid distortion of our view is caused by the turbulent atmosphere having numerous pockets and layers of air with varying refractive indices. As the starlight passes through these different patches they make the light bend slightly in different directions and the star appears to jiggle around. Professional observatories try to minimise this problem by putting the telescope on the top of a mountain, above most of the atmosphere, or out in space where there is no atmosphere at all to get in the way.

Talking about the atmosphere reminds me of one evening when I was working at the Ayers Rock Resort in Central Australia. We were there to provide entertainment and the opportunity for people to find out a little more about the night sky. It didn't matter whether the seeing was good or bad; we just wanted to give people the chance to look through a telescope. One night a person asked one of the most pointless questions I have ever heard. They wanted to show off that they knew a bit about astronomy, so they asked 'How often do you get sub-arcsecond seeing?'. This was a question about how steady the atmosphere was, which determines how much detail (or resolution) you can see. But sub-arcsecond seeing is excessively small criteria to meet. It is a question you would ask at the best viewing locations, with the steadiest atmosphere on the planet (where the best telescopes are found for that reason), not in the middle of a desert at an altitude of only 500 metres above sea level. Conditions at that level of clarity did occasionally occur, but they existed for only ever so brief a moment. I still smile when I think about that question.

What the atmosphere carries can also be a problem. Living in Central

Australia, one of the biggest threats was dust, bright red dust. It got everywhere, including on the telescope optics, no matter how careful we were. Massive dust storms in The Centre were frequent and we often had to race to cover the telescopes as best we could. Dust on the optics diminishes the effectiveness of your telescope, so you need to keep them as clean as possible. Even worse though is scratching the optics when trying to remove the dust. As annoying as it can be, it is often better to leave some dust on the optics, providing it isn't too bad. But if you must clean it, you need to be very, very careful. Better still, give it to a professional to clean.

And it wasn't only dust that created problems for our telescopes at the Ayers Rock Resort. We had to cope with the airborne grease from the regular barbeques held 500 metres away. As a telescope owner, that shows you have to be forever vigilant as even things you think couldn't possibly affect your telescope have to be carefully monitored.

Dew is another atmospheric problem that is the bane of all telescopes. The need to protect telescope optics from dew has spawned numerous accessories to try and cope with the problem. Dew caps, hairdryers and small heaters have been employed. The best protection against dew, however, is not to use the telescope when dew is a problem. Of course, that's not always possible, and you have to cope with the problem the best way you can. One word of advice though, never wipe any dew off the optics. This will only scratch the optics and degrade the view through your telescope.

Everyone who has a telescope has had to deal with dew, but the most notable instance I have come across was in Central Australia. Generally, the air out there is dry, one per cent humidity during Winter and 10 per cent humidity in Summer was the norm. Dew was never a problem, except, for some reason, as the seasons transitioned from Summer to Winter. At this time there were two weeks where the dew was so heavy it looked as if we had hosed down the telescopes. Standing outside for just five minutes rendered us dripping wet. It is the only place I've been where I had to wear a waterproof jacket to keep dry on a crystal clear night. After two weeks, it would suddenly disappear. The same thing would then happen as we transitioned from Winter back to Summer.

A lot of people are initially disappointed when they first look through a telescope. They expect to see large and colourful objects taking up the entire view. Unfortunately, this is one of the downsides from the proliferation of spectacular images we see everywhere. These images were taken through telescopes using high quality cameras, often with exposures that lasted minutes to hours. And a lot of the images were originally quite small, but because of their exceptional resolution they can be blown up to a large size without losing much detail. But, and it's a big but, when you look through a telescope and realise you're not going to see a big, bright object you will be amazed at the detail you can see. With a bit of patience and an appreciation of just what it is you are looking at, you will remember your view of the night sky through your telescope forever.

At last, we come to the end of our journey discovering the universe. I do, however, have one final piece of advice. Be curious, be patient, and be persistent. What you are doing is often difficult to master in the beginning, but with practice, what started out hard will become easy. And once that happens, you can get the most from your time under the stars.

Afterword

I wrote this book not to give endless facts and figures (although some do sneak in), but more as a means to share some of the lessons I have learnt, mistakes I have made, and knowledge I have gained during a lifetime of presenting astronomy and science to the public.

During the course of reading about my journey, I hope you have discovered some new things about our universe and picked up some tips on how to share this passion with others. Even the smallest of things and throw-away moments of encouragement can have a profound impact on someone's life and career choices.

Astronomy, and curiosity in general, doesn't have to be complicated. Sometimes the simplest things can give us a deep insight into the universe we inhabit if you are ready to stop and look.

My astronomical journey is far from over, and I have enjoyed sharing my journey so far with you. I hope that I have helped to inspire you to pursue your own life in science and to encourage the next person in theirs.

And always remember, wonder can be found anywhere and everywhere you look in our vast universe.

Images, Diagrams and Tables

A close-up of the 25 cm astrograph on the main deck of the observatory.
One of the signs at the start of a path leading to our observatory at the Ayers Rock Resort.
The same sign is now hanging in my backyard.
Solar viewing on the main deck with our Japanese guide.
Lightning over a distant Uluru.
Water cascading off Uluru during a storm.
Moonrise over Kata Tjuta.
Dune with red sand and spinifex on the western edge of the Simpson Desert at Old Andado Station, east of Alice Springs.
Remains of the old Ghan Railway track in South Australia.
Moonrise over a glowing, red Uluru.
According to this edition of Cleo magazine, the observatory was Number 1 in the Romantic Getaway list.
Country Style magazine article about the observatory.
Article about the observatory in the Japanese astronomy magazine Tenmon Guide.
Me being interviewed by a TV crew not long after we started.
Waiting to be interviewed in the early morning sub-zero temperature on top of a dune.

Chapter 4 - Return to Sydney

View of the Sydney Harbour Bridge from the Southern Dome of Sydney Observatory.
View of city buildings from the Northern Dome of Sydney Observatory.

Chapter 5 - Comet Shoemaker-Levy 9

Standing beside my telescope showing visitors the Comet Shoemaker-Levy 9 impacts on Jupiter.

Chapter 6 - Expanding horizons

Our portable planetarium set up in a school library.
Domes of the Macquarie University Observatory.

Chapter 7 - A move to Orange

One of the first site options offered as a potential location for the planetarium.
An artist impression of designs for the planetarium facility at the Orange Botanic Gardens.
An artist impression of the interior of the Botanic Gardens design.
The International Year of Astronomy poster advertising our talk on telescopes.
Evening viewing during the Slow Summer festival.
Orange Planetarium Incorporated display at the Orange Rotary Expo.
Artist impression of the exterior for the new Orange Regional Conservatorium and Planetarium facility.
The site for the new Orange Regional Conservatorium and Planetarium facility.

PART TWO - LESSONS LEARNT, KNOWLEDGE GAINED

Chapter 17 - Constellations

The Saucepan. Image courtesy of Geoffrey Wyatt.

Chapter 18 - Further afield

The Large Magellanic Cloud galaxy.
The Small Magellanic Cloud galaxy and the globular cluster 47 Tucana.
The Southern Milky Way and its dust clouds.
The location of the Large and Small Magellanic Clouds relative to the Southern Cross.
Crux and the Magellanic Clouds. Image courtesy of Geoffrey Wyatt.
M42, The Great Nebula in Orion. Image courtesy of Diego Colonnello from Sidereal Trading.

PART THREE - TELESCOPES AND STUFF

Chapter 21 - How do optical telescopes work?

A single biconvex lens focusses light passing through it to a focal plane.
When viewed at the focal point of a lens the image appears upside down.
The front, primary lens brings the light to a point, and the second lens magnifies the generated image.
A parabolic mirror does the same job as a biconvex lens and focusses light that strikes it to a focal point.

Chapter 22 - Telescope bits and pieces

A 280 mm telescope and an 80 mm telescope.
The focal length of a lens is the distance from the lens to the focal point.
A refracting telescope has one lens at the front of a tube and a second lens to look through at the other end.
An 80mm refracting telescope.
A Newtonian telescope on a Dobsonian mount.
A Newtonian telescope uses a parabolic mirror at the bottom to focus the light back up the tube before reflecting it out to the side and a lens for viewing.
The light in an SCT first passes through a correcting lens to adjust for using a spherical mirror instead of a parabolic mirror.
A Schmidt Cassegrain telescope (SCT) is shorter than a Newtonian telescope but still uses a mirror to focus the light.
The Dobsonian is a variety of Alt-Azimuth mount where it swings the telescope up and down, left and right.
The German Equatorial mount swings the telescope around an axis aligned to the South Celestial Pole.
Different sizes and styles of eyepieces.
A finderscope is used to point the telescope at the observed object.

Chapter 23 - Buying a telescope

A proud owner and his first homemade telescope.

My first telescope 40 years later.

Carrying a large telescope any distance is difficult. I have lots of problems moving this one around, especially to remote locations like desert paths.

About the Author

Rod Somerville is a newspaper columnist, radio presenter, guest speaker and walking astronomical encyclopedia. What began in an outer suburb of Sydney as an obsession with science fiction at the age of 10, morphed into a life of showing the universe to anyone who would stop and listen all around Australia. With degrees in Physics and Mathematics, Rod has worked as a science educator for over 40 years. He established and ran a public observatory at Uluru in Central Australia for several years and since returning to the east coast has been actively involved in the public education of science wherever he has gone. Rod is currently the President of Orange Planetarium Incorporated where he is in the process of establishing a planetarium in his adopted town of Orange, NSW. Visit www.southernskies.com.au to keep up-to-date on his latest endeavours.

www.ingramcontent.com/pod-product-compliance
Ingram Content Group UK Ltd.
Pitfield, Milton Keynes, MK11 3LW, UK
UKHW020658041026
15590UKWH00001B/339

9 780645 098709